Programming FPGAs

Getting Started with Verilog

Simon Monk

New York   Chicago   San Francisco
Athens   London   Madrid
Mexico City   Milan   New Delhi
Singapore   Sydney   Toronto

Cataloging-in-Publication Data is on file with the Library of Congress

McGraw-Hill Education books are available at special quantity discounts to use as premiums and sales promotions or for use in corporate training programs. To contact a representative, please visit the Contact Us page at www.mheducation.com.

Programming FPGAs: Getting Started with Verilog

1 2 3 4 5 6 7 8 9   DOC   21 20 19 18 17 16

ISBN   978-1-25-964376-7
MHID   1-25-964376-X

Sponsoring Editor	**Project Manager**	**Indexer**
Michael McCabe	Patricia Wallenburg,	Claire Splan
Editorial Supervisor	TypeWriting	**Art Director, Cover**
Donna M. Martone	**Copy Editor**	Jeff Weeks
Production Supervisor	James Madru	**Composition**
Lynn M. Messina	**Proofreader**	TypeWriting
Acquisitions Coordinator	Claire Splan	
Lauren Rogers		

To my wife, Linda

About the Author

Dr. Simon Monk has a bachelor's degree in cybernetics and computer science and a Ph.D. in software engineering. He is now a full-time writer and has authored numerous books, including *Programming Arduino, Programming Raspberry Pi, Hacking Electronics,* and is co-author of *Practical Electronics for Inventors.* Dr. Monk also runs the website MonkMakes.com, which features his own products. You can follow him on Twitter, where he is @simonmonk2.

CONTENTS

PREFACE

Wouldn't it be great to be able to have your own chips made that do exactly what you want them to? Well, field-programmable gate arrays (FPGAs) let you get pretty close to this ideal. Rather than being a chip specifically designed for you, an FPGA is a general-purpose chip that can be configured to act any way you want.

What is more, the way you configure your FPGA is either by drawing schematics or using a hardware-definition language that could (if your design is successful) also be used to manufacture chips that are actually custom chips to your design. The language in question is called *Verilog*, and although this book shows you how to make designs using a schematic editor, the main focus is on learning the Verilog language.

The FPGA configuration can be changed as many times as you like, making it a great tool for prototyping your designs. If a problem rears its head, you can just reprogram the device until you get all the bugs out. The ultimate mind-bending flexibility comes when you realize that you could actually configure your FPGA to include a processor capable of running programs.

In this book you will learn both the general principals of using FPGAs and how to get the examples described in this book up and running on three of the most popular FPGA evaluation boards: the Mojo, the Papilio One, and the Elbert 2.

Although, logically, a microcontroller can do pretty much anything that a FPGA can, a FPGA generally will run faster, and some people find it easier to write a description of logic gates and hardware than they do a complex algorithm. You can use a FPGA to implement a microcontroller or other processor (and people do).

Perhaps the most compelling reasons to try out programming some Verilog on one of the many low-cost FPGA boards is simply to learn something new and have some fun!

ACKNOWLEDGMENTS

As always, I thank Linda for her patience and support.

At TAB/McGraw-Hill and MPS Limited, my thanks go out to Michael McCabe, Patty Wallenburg, and their colleagues. As always, it was a pleasure working with such a great team.

Thanks also to Duncan Amos for his helpful and thorough technical review.

1

Logic

Field-programmable gate arrays (FPGAs) are digital devices that rely on digital logic. Computer hardware uses digital logic. Every calculation, every pixel rendered onto a screen, and every note of a music track is created using the building blocks of digital logic.

Although digital logic at times looks more like an abstract mathematical concept rather than physical electronics, the logic gates and other components of digital logic are constructed from transistors etched onto integrated circuits (ICs). In the case of FPGAs, a circuit can be designed by drawing logic gates that are then mapped onto general-purpose gates on the FPGA that are connected together so as to implement your logic design. Alternatively, the logic can be described using the Verilog (or another) hardware description language.

You can still buy chips that contain a small number of logic gates, such as the 7400, which has four two-input NAND gates. However, these are really only to maintain aged systems that use the chips or for educational use.

Logic Gates

Logic gates have inputs and outputs. These digital inputs and outputs can be either *high* or *low*. A *low* digital input or output is indicated by a voltage close to 0 V (ground). A *high* digital input is usually something over half the supply voltage of the logic, and a *high* digital output is at the positive supply voltage. The supply voltage for a FPGA is usually one of 1.8, 3.3, or 5 V, and most FPGAs can operate over a range of voltages. Some allow multiple logic voltages to be used on one device.

Describing logic gates can be tricky because the names of the logic gates (not, and, or, etc.) also mean something in English too. To avoid any confusion, I have capitalized the gate names.

The NOT Gate

The simplest logic gate is the NOT gate (sometimes called an *inverter*). It has a single input and a single output. If the input is high, the output will be low, and vice versa. Figure 1-1 shows the schematic symbol for a NOT gate. The truth table lists every possible combination of input and the resulting output. By convention, inputs tend to start at the beginning of the alphabet with names such as A, B, and C. Outputs are quite often called Q or letters near the end of the alphabet such as X, Y, and Z.

Figure 1-1 *A NOT gate.*

To describe how logic gates or groups of logic gates behave, you can use something called a *truth table.* This specifies the output that the logic will provide for every possible combination of inputs or outputs. For a NOT gate, this is shown in Table 1-1. The letters H and L or the numbers 1 and 0 are used in place of *high* and *low.*

Input	Output
L	H
H	L

Table 1-1 *Truth Table for a NOT Gate*

If you were to connect the output of one NOT gate to a second NOT gate, as shown in Figure 1-2, the output of the combination would always be the same as the input.

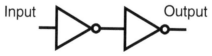

Figure 1-2 *Two NOT gates.*

The AND Gate

As its name suggests, the output of an AND gate will be high only if all its inputs are high. Figure 1-3 shows the symbol for a two-input AND gate, and Table 1-2 is the truth table for the AND gate.

Figure 1-3   *An AND gate.*

Input A	Input B	Output Q
L	L	L
L	H	L
H	L	L
H	H	H

Table 1-2   *Truth Table for an AND Gate*

The OR Gate

The OR gate also behaves as its English-language name would suggest. The output of an OR gate will be high if any of its inputs are high. Figure 1-4 shows the symbol for a two-input OR gate, and Table 1-3 is the truth table for the OR gate.

Figure 1-4   *An OR gate.*

Input A	Input B	Output Q
L	L	L
L	H	H
H	L	H
H	H	H

Table 1-3   *Truth Table for an OR Gate*

NAND and NOR Gates

The little circle on the output of the NOT gate in Figure 1-1 is what indicates the inverting function of the gate. A NAND gate (NOT AND) is an AND gate with an inverted output, and a NOR gate is an OR gate with an inverted output. Figure 1-5 shows the symbols for the two gates, and Tables 1-4 and 1-5 are the truth tables for the two gates.

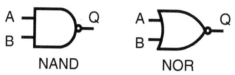

Figure 1-5   *NAND and NOR gates.*

Input A	Input B	Output Q
L	L	H
L	H	H
H	L	H
H	H	L

Table 1-4   *Truth Table for a NAND Gate*

Input A	Input B	Output Q
L	L	H
L	H	L
H	L	L
H	H	L

Table 1-5   *Truth Table for a NOR Gate*

Both NAND and NOR gates are described as *universal gates* because either of them can be used to make any other type of gate by inverting either the inputs to the gate or the outputs from the gate. What's more, you can use either a NAND or an OR gate to make a NOT gate simply by tying together the two inputs. For example, Figure 1-6 shows how you can make an AND gate using three NOR gates.

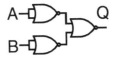

Figure 1-6 *Making an AND gate using NOR gates.*

De Morgan's Law

The design of Figure 1-6 makes use of a law of logic called *De Morgan's law*. This is best explained as follows: inverting the result of two inputs "anded" together gives the same result as inverting both the inputs and "oring" them together. In Figure 1-6, what we actually have is both the inputs being inverted by a NOT gate (NOR gate with inputs tied together) "ored" together, and then the output of the OR gate is itself inverted, giving the overall result of "anding" all the inputs. This is a useful trick to know.

You can check how this works by examining the truth table in Table 1-6, which includes the intermediate states of A and B, the result of inverting A and B, respectively.

Input A	Input B	NOT A	NOT B	Output Q
L	L	H	H	L
L	H	H	L	L
H	L	L	H	L
H	H	L	L	H

Table 1-6 *Truth Table for Making an AND Gate from NOR Gates*

XOR Gates

The OR gate discussed earlier is an inclusive or; that is, A or B can be high or both can be high for the output to be high. When we say *or* in English, we tend to mean the exclusive version of or. Do you want cream *or* ice cream with your dessert—the implication being that the attractive option of having both is not allowed. This type of OR in a logic gate is called an *exclusive OR*, shortened to XOR, and is very useful because it allows inputs to be compared. The output of an XOR will be high as long as the inputs are the same, irrespective of whether the inputs are high or low.

Figure 1-7 shows how you can make an XOR gate using four NAND gates. Because XOR gates are used quite commonly, the symbol for an XOR gate is shown alongside the NAND gate version. Table 1-7 shows the truth table for an XOR gate.

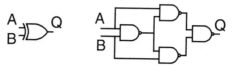

Figure 1-7   *Making an XOR gate from four NAND gates.*

Input A	Input B	Output Q
L	L	L
L	H	H
H	L	H
H	H	L

Table 1-7   *Truth Table for an XOR Gate*

Binary

If instead of thinking of the inputs and outputs to logic gates as being voltages of high or low, we can think of them as numbers (1 for high and 0 for low), and then we can start to get an inkling of how a computer could make use of logic gates to manipulate numbers. However, a range of numbers of just 0 and 1 is not going to be that useful.

We use *number base 10*; that is, we use 10 symbols to represent our numbers 0, 1, 2, 3, 4, 5, 6, 7, 8, and 9. This is done probably because of our having 10 fingers. When we need to represent a number larger than 9, we have to use two symbols—10, 45, 99, and so on. The right-most digit (called the *least significant digit*) is the number of units, the next digit to the left is the number of tens, the next the number of hundreds, and so on.

If we had decided to use the number of noses we have to count with rather than the number of fingers, then we would use *number base 2*. You either have 0 or 1. If you want to represent a number bigger than 1, then you must have more than one digit. Number base 2 is also called *binary*, and one binary digit is called a *bit*. Generally, to do something useful, we need a few bits grouped together, just like we need a group of decimal digits for larger numbers.

Table 1-8 shows how the numbers 0 to 7 are represented using 3 bits. Note that when displaying binary numbers, you usually include the leading zeros.

Decimal Number	Binary Number
0	000
1	001
2	010
3	011
4	100
5	101
6	110
7	111

Table 1-8 *The Numbers 0 to 7 in Binary and Decimal*

If you break down a decimal number, say, 123, you could write that as $1 \times 100 + 2 \times 10 + 3$. You can use this approach if you want to know the decimal representation for a number written in binary. For example, 111 in binary is $1 \times 4 + 1 \times 2 + 1 = 7$ in decimal. If you pick a bigger binary number to convert to decimal, say, 100110, the decimal value is $1 \times 32 + 0 \times 16 + 0 \times 8 + 1 \times 4 + 1 \times 2 + 0 = 38$.

A *byte* is 8 bits, and with a byte, the decimal equivalents of the digit positions are 128, 64, 32, 16, 8, 4, 2, and 1. If you add all those together, this means that you can represent any decimal number between 0 and 255. Each time you add another bit, you double the number range that you can represent. The numbers get big pretty quickly. A modern computer will do arithmetic on 64 bits at a time, giving a range of numbers from 0 to nearly 18,000,000,000,000,000,000.

Adding with Logic

Logic gates allow you to do arithmetic on binary numbers. Because binary numbers are just numbers represented in bits, you can do arithmetic on any numbers using logic gates. Figure 1-8 shows how you can make a binary adder using logic gates.

Table 1-9 shows the truth table for the single-bit adder. This time, however, there are two outputs: the sum and the carry bit, which will be 1 when the result of the addition is too big to hold in 1 bit.

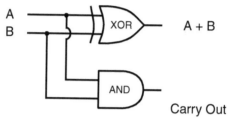

Figure 1-8   *A single-bit adder.*

Input A	Input B	Output Sum (A + B)	Output Carry-Out
0	0	0	0
0	1	1	0
1	0	1	0
1	1	0	1

Table 1-9   *Truth Table for a Single-Bit Adder*

Looking at the table, you can see that if A and B are both 0, then the sum is 0. If one of them is 1, then the sum will be 1. However, if both A and B are 1, then the sum digit itself will be 0, but we will want to carry a 1 to the next number position. In binary, 1 + 1 is 0 carry 1 (or decimal 2).

This does mean that if we want to add more than 1 bit at a time, the next adding stage will have three inputs (A, B, and carry-in). This is a bit more complicated to accomplish because we have to add together three digits rather than two. Figure 1-9 shows an adder stage that takes a carry-in digit.

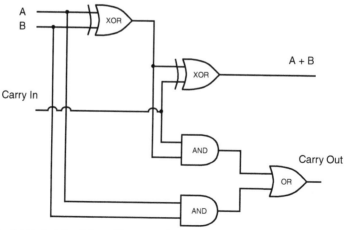

Figure 1-9   *A single-bit adder with a carry-in bit.*

You will never need to create an adder like this from multiple gates because adders are available as single components that you can use in your designs. However, it is interesting to see how they can be constructed from basic gates.

If we had eight of these stages, we could use them to add 2 bytes together. Every computer's central processing unit (CPU) will have a hardware adder made up of logic gates in a very similar way to this. A 32-bit processor will be able to process 32-bit numbers in one go, and a 64-bit machine will have 64 stages like Figure 1-9 to add 64 bits at a time.

Flip-Flops

In the preceding examples, the gates are arranged so that you will always get the same outputs for a given set of input values. If you start feeding back the outputs of logic gates to an earlier gate that affects the first gate's inputs, then you can give your network of logic gates the ability to remember things. This kind of logic is called *sequential logic*.

One of the fun things about electronics is that it is a relatively modern field, and its pioneers could ditch the Latin pomposity of earlier naming conventions and call things that flip and flop from one state to another a *flip-flop*. For those with a yearning to call them something less interesting, you may also refer to them as *bistables*.

Set-Reset Flip-Flop

Take a look at the schematic of Figure 1-10. This is called a *set-reset (SR) flip-flop*. Having introduced the idea that high means 1 and low means 0, let's stick to numbers for logic values. Imagine that initially S (set) is 1 and R (reset) is 0. Because one input (S) to NOR gate A is 1, then it doesn't matter about the other input to NOR gate *A*, the output will be 0. This 0 output from NOR gate A means that the top input to NOR gate B will also be 0. The other input to NOR gate B (R) is also 0, so the output of NOR gate B will be 1, making the lower input to NOR gate A also 1. Next, imagine that you now set S to 0. This will have no effect on the output of NOR gate A.

Having set the output Q to 1 by setting S to 1, Q will not change whatever we do with S: it will change only if R is set to 1, making the output of NOR gate B (Q) 0. The output of NOR gate A is labeled as Q with a line over it. The line (called a *bar*) indicates that the output is inverted, and $\bar{Q}$ will always be the inverse of Q.

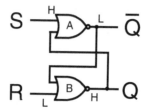

Figure 1-10 *A set-reset flip-flop.*

The SR flip-flop just described is used occasionally, but perhaps the most commonly used and flexible type of flip-flop is the *D flip-flop*. This can be made from a load of NAND or NOR gates, but you only ever need to use it as a logical block in its own right. Figure 1-11 shows the schematic symbol for a D flip-flop.

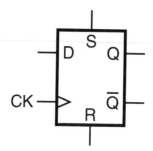

Figure 1-11 *A D flip-flop.*

The D flip-flop still has the S, R, Q, and $\bar{Q}$, but it has two extra pins D and CK (clock). The clock symbol is often shown as a little triangular notch. You can still use S and R to set and reset the flip-flop, but you are more likely to use D and CK.

The concept of a clock is something that is essential to digital electronics. It synchronizes the system so that the slight delays caused by logic gates changing from high to low and propagated through multiple paths between each other do not cause glitches if the outputs haven't finished settling. A clock generally will be connected to a signal that flips back and forth between high and low. FPGAs have a built-in clock signal that you will use in most of the examples in this book. This signal is 12 to 50 MHz depending on which of this book's example FPGA boards you are using. This is 12 to 50 million cycles of high/low per second.

When the clock signal goes high, whatever value of D (0 or 1) will be latched into the output Q. This may not sound very exciting, but it does allow you to construct the much-used shift registers and counters described in the following sections.

Shift Registers

Figure 1-12 shows four D flip-flops arranged with the Q output of one D flip-flop connected to the D input of the next. All the clocks for the D flip-flops are connected together. This arrangement is called a *serial-to-parallel shift register*.

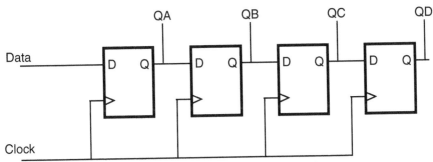

Figure 1-12   *A 4-bit serial-to-parallel shift register using D flip-flops.*

Imagine that clock and D are both connected to push switches, and QA to QD are all connected to light-emitting diodes (LEDs) that light when they are high. Hold down the D switch (to make it high), and then, when you briefly press the clock switch, the high of D will become latched into the first flip-flop. Release the D switch (to make it low), and when you momentarily press clock again, two things will happen simultaneously:

1. The value that was at QA (high) will be latched into the second flip-flop and will make QB high.

2. The current value of D (low) will be latched into the first flip-flop and will make QA low.

Each time you pulse clock high, the bit pattern will shuffle to the right along the flip-flops. Note that it is the transition of the clock from 0 to 1 that triggers the bits to shuffle to the right. You could, of course, add as many D flip-flops as you like.

Binary Counters

D flip-flops are very flexible components. Arranged with the inverted output $\bar{Q}$ from one flip-flop connected to the clock input of the next flip-flop (Figure 1-13), the flip-flops can be arranged to count in binary.

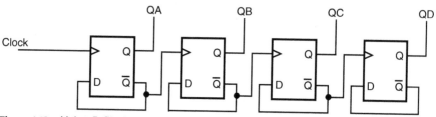

Figure 1-13 *Using D flip-flops as a counter divider.*

Each time that the clock signal is pulsed, flip-flop A toggles its state between 0 (low) and 1 (high). The output of the flip-flop then provides pulses at half the frequency to the next stage, and so on. This is why counters are also called *frequency dividers*. Thus, if the frequency at the clock is 24 MHz, the frequency at QA will be 12 MHz and so on down the line, halving each time.

Digital watches and clocks usually operate at a clock frequency of 32.768 kHz. This number is chosen because if divided by two 15 times, it results in a frequency of 1 Hz (1 pulse per second). In Chapters 3 and 4 you will implement this example of a counter on your FPGA board, first by drawing it as a schematic rather like Figure 1-13 and then by writing a few lines of Verilog code.

Summary

This chapter provides only a taste of the wide and complex field that is digital electronics. As the book progresses, you will meet other types of digital components.

Although it is important to understand the basics of digital electronics when using an FPGA, use of a language such as Verilog means that you can create complex designs without the need to understand exactly what is being created for you at gate level by the software. In Chapter 2, you will get to learn exactly what an FPGA is and meet the FPGA prototyping boards used in this book, the Elbert V2 and the Mojo and Papilio boards.

2

FPGAs

Now that you have looked at the basic building blocks of digital electronics, it's time to take a look at FPGAs to see exactly what they do and meet the Elbert 2, Mojo, and Papilio evaluation boards that we can then configure to act as a complex digital system.

How an FPGA Works

An FPGA is made up of general-purpose logic cells (that have 64 inputs and one output). When the FPGA is being configured, these general-purpose logic cells are connected together using yet more logic gates, and finally, some of the outputs can be connected to special general-purpose input-output (GPIO) cells that allow them to be used as digital inputs or outputs via the physical pins on the FPGA chip's package. If the evaluation board you are using has built-in LEDs and switches, then these will be permanently connected to certain GPIO logic cells of the FPGA.

The logic blocks that make up an FPGA use a lookup table (LUT). The lookup table will have a number of inputs, say, six inputs and a single output. Imagine it as a truth table with six inputs (64 combinations). The table can be configured to specify an output value (0 or 1) for every possible combination of the six inputs to the gate. The contents of these LUTs, combined with other routing information, are what give the FPGA its logic.

The LUT will often be combined with extra components such as a flip-flop to make an individual logic block. Figure 2-1 shows a logical view of how all this is arranged.

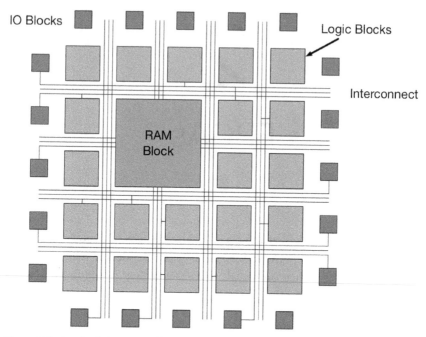

Figure 2-1 *Logical structure of an FPGA.*

GPIO pins on the FPGA chip are connected to special-purpose input-output (IO) blocks that provide buffered microcontroller-like inputs and outputs that typically can source or sink a few tens of milliamperes.

The vast bulk of the functional units in the FPGA will be logic blocks, and a typical modern FPGA may have from 200,000 to several million of these blocks. There also may be a fixed random access memory (RAM) block for use when the FPGA is to be configured as a processor or just needs to store a lot of data such as in the WAV sound file player example of Chapter 8. Taking this idea of some special-purpose areas of the FPGA to its extreme, you find high-end systems-on-a-chip (SoC) FPGAs that include fixed high-performance processor cores and memory on a chip that also includes configurable logic cells. FPGAs are also often used to prototype an application-specific IC (ASIC) for very large production runs.

Routing between such vast numbers of logic blocks is pretty tricky, but fortunately for us, we don't have to do it—that's what the design software is for.

The information in the LUTs and the routing matrix (that defines the interconnections) is volatile. When you lose power, all that information disappears, and the FPGA reverts to its original state. To configure the FPGA, the configuration is usually stored outside the FPGA in electrically erasable programmable read-only memory (EEPROM) or Flash memory (which is not lost when the power is removed). The FPGA generally will have a fixed-hardware loading interface built into it that will pull in the configuration as the FPGA starts up. This typically takes less than 1/5 of a second.

The Elbert 2

Figure 2-2 shows the Elbert 2 board. This board is an FPGA evaluation board. Its sole purpose is to allow you to experiment with FPGAs and learn how to use them without having to spend a load of time soldering or designing peripheral electronics to connect to it because it's all already there.

These boards are available directly from Numato Laboratories (http://numato .com) or from Amazon.com and various other sources. The price of the entire board was just $29.95 at the time of this writing. Because the Elbert 2 has a selection of built-in buttons and LEDs, the only other thing you will need is a

Figure 2-2 *The Elbert 2–Spartan 3A FPGA development board.*

USB-to-mini-USB lead. All the example boards used in this book are quite happy being powered from the USB port.

The board has the following features:

- Spartan XC3S50A FPGA in TQG144 package (1584 logic cells, 54 kbits of RAM)
- 16-MB SPI Flash memory for configuration
- USB 2.0 interface for onboard Flash programming
- 8 LEDs
- 6 push buttons
- Eight-way dual in-line package (DIP) switch
- VGA connector
- Stereo jack
- Micro secure digital (SD) card adapter
- 3 seven-segment LED displays
- 39 IOs for you to use
- Onboard voltage regulators

The Mojo

The Mojo (Figure 2-3) is another popular FPGA evaluation board. It's a much more elegant and compact board than the Elbert 2. While it does have a row of built-in LEDs, to really experiment, you should also get a Mojo IO Shield. This adds a four-digit LED display and a whole load of push and slide switches. The Mojo uses a faster and more powerful FPGA chip than the Elbert.

The Mojo has the following features:

- Spartan 6 XC6SLX9 FPGA (9152 logic cells, 576 kbits of RAM)
- 84 digital IO pins
- 8 analog inputs
- 8 general-purpose LEDs
- Reset button
- LED to show when the FPGA is correctly configured

Figure 2-3 *The Mojo board (top) and with IO Shield attached (bottom).*

- Onboard voltage regulation supply voltage 4.8 to 12 V or USB
- A microcontroller (ATMega32U4) used for configuring the FPGA, USB communications, and reading the analog pins
- Arduino-compatible bootloader allowing you to easily program the microcontroller as well as the FPGA
- Onboard Flash memory to store the FPGA configuration file

The Mojo IO Shield adds the following features to the Mojo:

- 24 LEDs
- 24 slide switches
- 5 push buttons
- 4 seven-segment LED displays

The Papilio

The Papilio One (Figure 2-4) does not itself have any built-in LEDs or switches. Like the Mojo, to experiment with it, it is designed to accept an interface board that just plugs onto the top of it. The one chosen for this book is the LogicStart MegaWing.

The Papilio One is available in 250- and 500-k models. The number refers to the FPGA part number of XC3S250E or XC3S500E. The other specifications of the Papilio One 500k are

- Xilinx Spartan 3E (10,476 logic cells, 74 kbits of RAM)
- 4 Mbits of SPI Flash memory
- USB connection
- 4 independent power rails at 5, 3.3, 2.5, and 1.2 V
- Power supplied by a power connector or USB
- Input voltage (recommended) of 6.5 to 15 V
- 48 I/O lines

The LogicStart MegaWing has the following features:

- 7-segment display, 4-character
- VGA port

Figure 2-4   *The Papilio One board (top) with LogicStart MegaWing (bottom).*

- Mono audio jack, ⅛ inch
- Micro joystick, five directions
- 8 analog inputs (SPI ADC, 12-bit, 1 Msps)
- 8 LEDs
- 8 slide switches, user input

Software Setup

All three evaluation boards use FPGAs from the manufacturer Xilinx, so they can all be programmed using the same Integrated Synthesis Environment (ISE) software. This will take you as far as generating a .bit file that then has to be transferred onto your evaluation board using a software utility specific to that board. This is pretty much a matter of choosing the file, selecting the serial port to which the board is connected, and then pressing Go.

Installing ISE

The design tools of the FPGA manufacturers are frankly bloated monsters. The ISE design tool is a 7-GB (that's right, gigabyte!) download. In many ways, getting and installing the design tool constitute the most time-consuming part of getting started with FPGAs.

The first step in obtaining ISE is to visit Xilinx.com with your web browser and find the ISE download page, which you will find by following the links: Developer Zone → ISE Design Suite → Downloads. Scroll down the "Downloads" area to ISE Design Suite (we used version 14.7), and select the option "Full Installer for Windows." Do not be tempted to download the newer Vivado Design Suite. This is only for newer Xilinx FPGAs and does not support the Spartan 3s used on the Elbert 2 and Papilio One.

There are Windows and Linux versions of the tool. In this book, we will just describe the process of getting up and running in Windows.

When you click on the "Download" button, a survey will appear that you have to complete, followed by a second registration form. Persevere and eventually the download will start, and you can go and do something else for a few hours while the download completes.

The downloaded file has a .tar extension that is not a normal Windows file format, so you will need to extract the contents of the tar file using a utility such as 7-zip (7-zip.org). With 7-zip installed, right-click on the .tar file, and select the option "Extract." Once the file is extracted, you will see a folder called something like Xlinx_ISE_OS_Win_14.7_1015_1. Inside this folder you will find the program xsetup.exe. Run xsetup.exe to install ISE. When you get to the window shown in Figure 2-5, select the option "Web Pack."

The main installer will start several subinstallers for other packages and ask you to agree to several license agreements. You can accept all the defaults offered to you for the remainder of the installation.

At the end of the installation process, select the option "Get Free Vivado/ISE WebPack License." You'll have to log in again and check a short form and then select "Vivado Design Suite: HL Web Pack ..." from the list of licenses available to you (Figure 2-6). Click on the button "Generate Node-Locked License." This will be e-mailed to you as an attachment called xilinx.lic; save the file, and then add the license from the License Configuration Manager that should still be open. If the License Configuration Manager is not open, you can open it from the "Manage

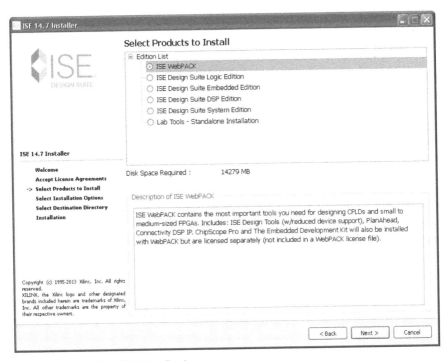

Figure 2-5 *Installing ISE Web Pack.*

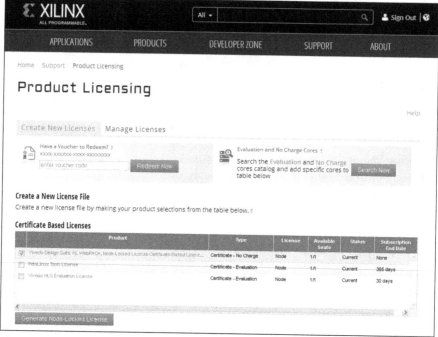

Figure 2-6   *Choosing an ISE license.*

License" option on the Help menu of Xilinx Platform Studio. The installation process will add a new shortcut on your desktop called "ISE Design Suit 14.7."

Installing Elbert Software

The Elbert board has a software utility for programming the board. This handles just the final step of copying the binary file generated by ISE onto the Elbert V2's Flash memory. There is also a USB driver to install for Windows users.

To set up your computer to use the Elbert, visit the product page for the Elbert V2 at numato.com (http://numato.com/elbert-v2-spartan-3a-fpga-development -board/), and click on the Downloads tab. You will need to download

- Configuration tool (used to program the board)
- Numato Lab USB CDC driver
- User manual

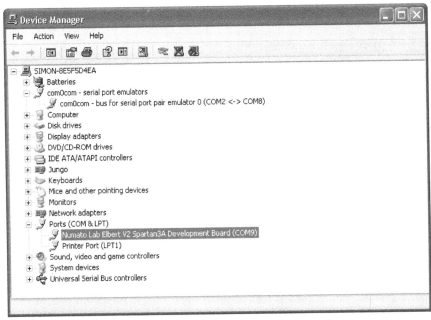

Figure 2-7   *Finding the COM port connected to the Elbert.*

To install the USB driver on Windows, plug the Elbert V2 board into your computer, and the New Hardware Wizard should start. Point the wizard at the extracted "numatocdcdriver" folder, and the driver should install, and the new hardware should be recognized.

After this, the Elbert V2 will be connected to one of the virtual COM ports of your PC. To find out which port, open the Windows Device Manager, and you should find it listed in the Ports section, as shown in Figure 2-7.

Installing Mojo Software

The Mojo board also has some software that you need to install. Visit https://embeddedmicro.com/tutorials/mojo-software-and-updates/installing-mojo-loader to download the Mojo loader for your computer's architecture and operating system.

Note that the Mojo board also has support for the Arduino IDE because the board also includes an ATMega microcontroller (as found in the popular Arduino microcontroller boards). In this book, you will not be programming this micro-controller or using the Arduino IDE to upload images to the board, so you will need the Mojo loader to upload onto the Mojo board.

Installing the driver for the Mojo is very similar to installing the driver for the Elbert 2. From the Found New Hardware driver, locate the driver from the instal-lation folder for the Mojo Loader (/mojo loader/driver).

Installing Papilio Software

You will find the loader for the Papillio at http://forum.gadgetfactory.net/index .php?/files/file/10-papilio-loader-gui/. The installer will automatically install the USB drivers for the Papillio and the loader. When you plug the board in, it will show up as FTDI Dual RS232.

Project Files

The files for all the projects in this book are contained in a GitHub repository at https://github.com/simonmonk/prog_fpgas. If you are already familiar with using git, then you can clone the code onto your own machine. If you are new to GitHub, then the easiest way to get the files onto your computer without having to install git is to navigate to the webpage in the preceding paragraph and then click on the "Download Zip File" button on the bottom right (Figure 2-8).

Figure 2-8   *Downloading a Zip archive of the project files.*

Unzip the files to a convenient location (say your desktop). The result will be a folder called prog_fpgas-master that itself contains several folders and files, including folders called elbert2, mojo, and papilio. Each of these folders for the three example boards contains project folders for all the project code for that particular board. Each project is prefaced by the number of the chapter in which it appears.

Much of the Verilog code is identical across the three boards, except where there are differences in the hardware (e.g., three-digit LED display digits rather than four in the case of the Elbert 2 or where differing clock frequencies of the boards have an effect).

Each project folder contains a file with the extension .xise. This is the project file, and double-clicking on it will start ISE on that project. The .bit file is a ready-made file for installation directly onto your FPGA board using the loader software for whichever board you have. Finally, the "src" folder contains the actual Verilog or schematic files for the project if you want to synthesize the design yourself to rebuild the .bit file.

The project folders start out pretty empty, but as soon as you open a project, you will find that ISE will add some files. When you actually synthesize a design and generate the .bit file, ISE will create even more files.

Summary

In this chapter, you have learned something about FPGAs and will now have your computer set up and ready to start some FPGA programming. Although this book is primarily about programming FPGAs using Verilog, before jumping into Verilog, Chapter 3 shows you how to program your FPGA board by drawing logic diagrams using gates and flip-flops.

3

Drawing Logic

The ISE design tool gives you two ways of programming your FPGA. One is to draw a familiar logic diagram, and the second is to use a *hardware description language* (HDL) such as Verilog. We will start with the schematic approach, although seasoned FPGA designers nearly always use Verilog or its rival VHSIC Hardware Description Language (VHDL).

Even if you don't plan to draw your schematics and just want to use Verilog, you should still work through this chapter because it explains how to use the ISE development tool as well as introducing concepts such as *user constraints files* that are common to Verilog and schematic design.

In this chapter, we will go into quite a lot of detail on using the ISE tool to get you up and running with what can be a quite daunting tool. You do not need to have your board connected to your computer until you have finished your design and are ready to program your FPGA.

A Data Selector Example

The first example that we will make is a data selector (Figure 3-1). The steps involved in this are mostly the same whichever FPGA board you are using. We will highlight where you need to do something different for your board.

This design has inputs A and B and an output Q. The output Q will either have the value of A or the value of B depending on SEL (select), which switches the output between the two inputs A and B. Table 3-1 shows the truth table for the selector. Note that the NOT gates are labeled INV (for inverter). This is another name for a NOT gate. An X in a truth table indicates that it does not matter whether that input is high or low; the output will not be affected by it.

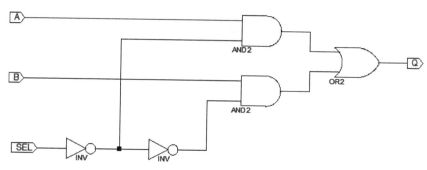

Figure 3-1 *A simple data selector.*

Input A	Input B	Input SEL	Output Q
L	X	L	L
H	X	L	H
X	L	H	L
X	H	H	H

Table 3-1 *Truth Table for an AND Gate*

The three inputs (A, B, and SEL) for this circuit will be hooked up to three of the push buttons on your FPGA board, and the output will be connected to one of the LEDs so that you can actually see the circuit in use.

You can either download the project files from GitHub, as described in Chapter 2, or follow the instructions below to create the projects from scratch. The data selector project is in folder ch03_data_selector.

Step 1: Create a New Project

The first step is to fire up ISE and then select File → New Project from the menu. This will open the New Project Wizard shown in Figure 3-2.

Enter "data_selector" in the Name field. In the Location field, navigate to the folder where you want to keep your ISE designs. The Working Directory field will automatically update to match this directory, so you don't need to change the Working Directory field.

Change the "Top-Level Source Type" drop-down to "Schematic," and then click "Next."

This will take you to the Project Settings shown in Figure 3-3.

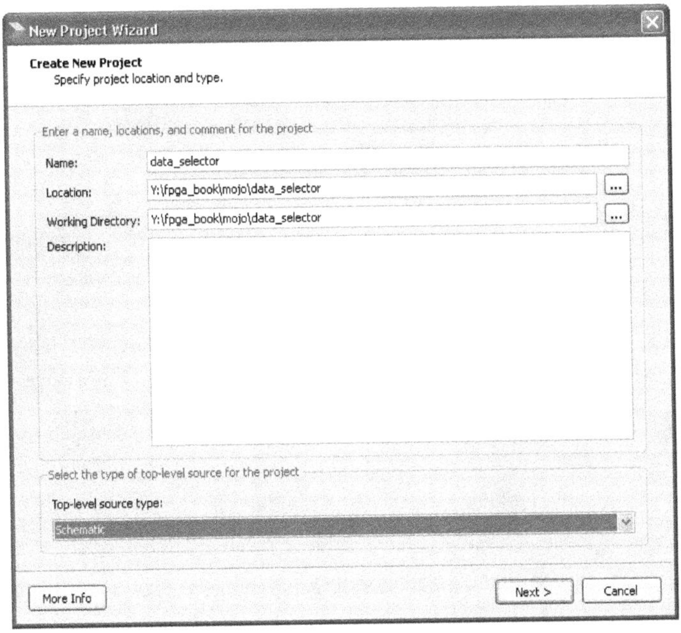

Figure 3-2 *The New Project Wizard in ISE.*

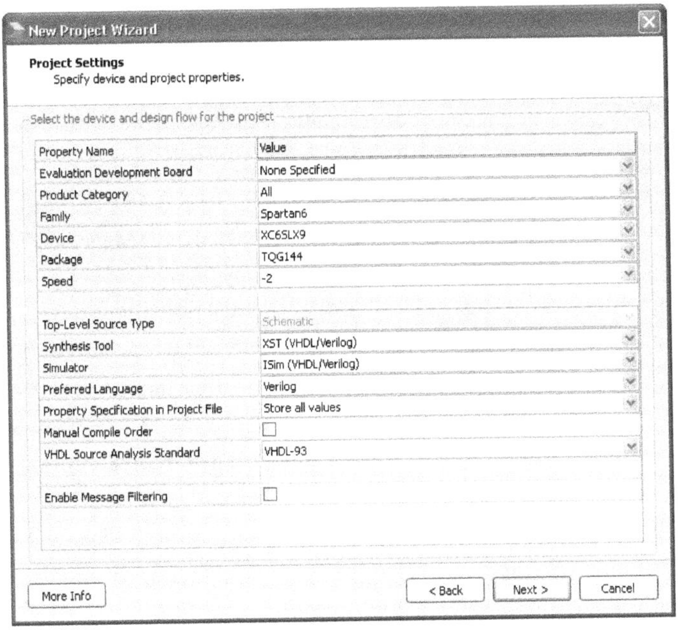

Figure 3-3 *The New Project Wizard—Project Settings (Mojo).*

The settings here will depend on which board you are using. Figure 3-3 shows the settings for the Mojo board.

Use Table 3-2 to make the settings, and then click "Next" again. You can find out more about these settings and other things specific to FPGA boards in Appendices B, C, and D.

Setting	Elbert 2	Mojo	Papilio One
Evaluation development board	None specified	None specified	None specified
Product category	All	All	All
Family	Spartan3A and Spartan3AN	Spartan6	Spartan3E
Device	XC3S50A	XC6SLX9	XC3S250E or XC3S500 (see note)
Package	TQ144	TQG144	VQ100
Speed	−4	−2	−4

Note: If you have a Papilio One 250k, use XC3S250E; for a Papilio One 500k, use XC3S500.

Table 3-2 *FPGA Board New Project Settings*

The Wizard will then show you a summary of the new project, and you can then click "Finish." This will create the new but empty project shown in Figure 3-4.

The screen is divided into four main areas. On the top left, you have the Project View. This is where you can find the various files that make up the project. It is organized as a tree structure. Initially, there are two entries in this area. There is the entry that says "data_selector," and the second entry that has an automatically generated name (xc6slx9-2tqg144) based on the device type and package. The latter will eventually contain two files, the schematic drawing that we are about to create and an "implementation constraints" file that defines how the inputs and outputs in the schematic connect to the actual switches and LEDs on the FPGA board.

You can also double-click on "xc6slx9-2tqg144" (or whatever it's called in your project) to open the project properties. Thus, if you made a mistake setting the project properties using the New Project Wizard, you can always correct it by double-clicking on this.

To the left, beneath the Project View, is the Design View. This will eventually list useful actions that you can apply to your design, including generating the binary file for deployment on your FPGA board.

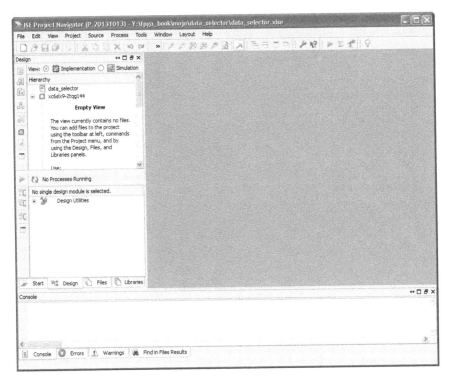

Figure 3-4 *A new project.*

The wide area at the bottom of the window is the Console. This is where error messages, warnings, and other information will appear.

The large area to the right of the window is the Editor area. When it comes to drawing the schematic, this is where you will do it.

Step 2: Create a New Schematic Source

To create a new schematic, right-click on "data_selector" in the Project View, and select the option "New Source…." This will open the New Source Wizard (Figure 3-5).

Select Source Type "Schematic," and enter "data_selector" in the File Name field. Then click on the three dots next to the Location field. To keep things tidy, create a new folder inside the default directory for the project called "src" (source), and then click "Next." A summary screen will appear, to which you can respond by clicking "Finish." This will result in a blank canvas being prepared for you in which

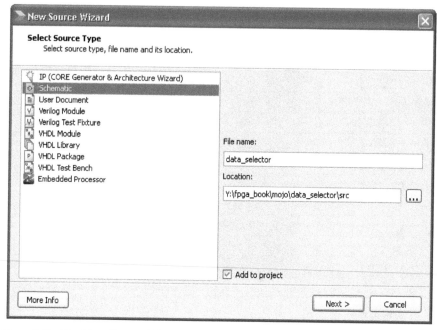

Figure 3-5 *The New Source Wizard.*

you can draw the schematic. This is shown in Figure 3-6 with the parts that you
are going to need to use labeled.

The icon menu bar running vertically to the left of the Editor area controls the
mode of the window and also what appears on the left-hand side of the window:

- The top icon (an arrow) puts the window into select mode. You will need to
  click on this before you can drag circuit symbols around or change their
  properties.

- Click on the "Add Wire" mode when you are connecting the gates and
  other circuit symbols together.

- IO markers are used to indicate the boundary between the schematic you
  are designing and the actual pins of the FPGA IC. This mode lets you add
  these symbols.

- Add logic symbols. This is the mode selected in Figure 3-6. The left-hand
  panel then divides into a top half that shows categories of circuit symbols
  and a bottom half that has a list of the component symbols in that category.

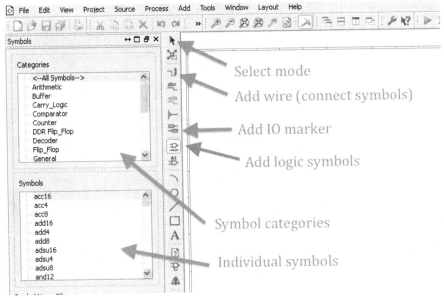

Figure 3-6   *The Schematic Editor.*

Step 3: Add the Logic Symbols

Put the screen into the Add Logic Symbols mode by clicking on the "Add Symbols" icon (see Figure 3-6). You are going to need to add two two-input AND gates, a two-input OR gate, and two NOT gates (inverters).

Click on the category "Logic", then select "and2" (two-input AND). Then click twice in the Editor area to drop the two AND gates. Then select "or2" and drop an OR gate in roughly the right location to the right of the AND gates. Finally, add in the two NOT gates ("inv" for inverter) below and to the left of the AND gates. Zoom in a bit (toolbar at the top of the window), and the Editor area should look something like Figure 3-7.

Step 4: Connect the Gates

Click on the "Add Wire" icon, and then connect the gates together in the arrangement shown in Figure 3-1. To make a connection, click on one of the square connection points, and drag it out to the connection point or line to which you want to connect. The software will automatically put bends in the line for you. If you need to do more things to tidy the diagram up, you can change to Select mode and drag the symbols and wires around. The end result should be Figure 3-8.

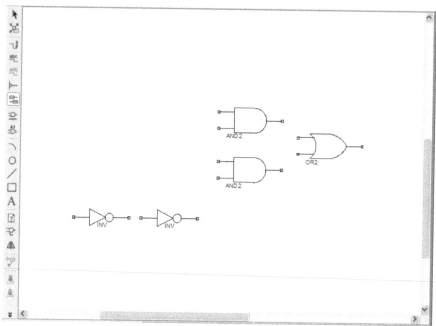

Figure 3-7   *The logic gates in position.*

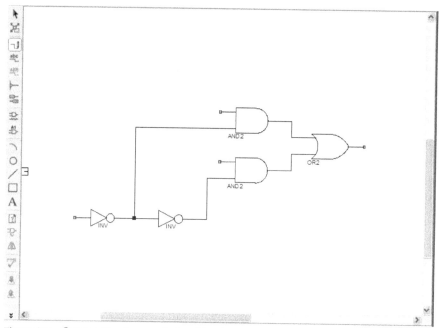

Figure 3-8   *Connecting the symbols with wires.*

Step 5: Add the IO Markers

Click on the "Add IO Markers" icon, and then add markers to all the inputs and the output by dragging the mouse out from the wire in question. Notice how the software figures out that the output is an output.

Initially, the connections are all given names like "XLXN_1" and so on. To change these names to something more meaningful, change to Select mode, right-click on an IO connector, and chose the menu option "Rename Port." Change the port names so that they agree with Figure 3-9.

Notice that we have called the output Q. Be careful when naming outputs because the word *OUT* is reserved for use by ISE, so you cannot call any of your connections OUT or you will get an error when you try to build the project, and the error message is a long way from "Don't call things OUT." You can find a list of other reserved names that you should not use here: www.xilinx.com/itp/xilinx10/isehelp/ite_r_verilog_reserved_words.htm.

The schematic is now complete, and now would be a good time to do File →
Save to save the schematic design to file.

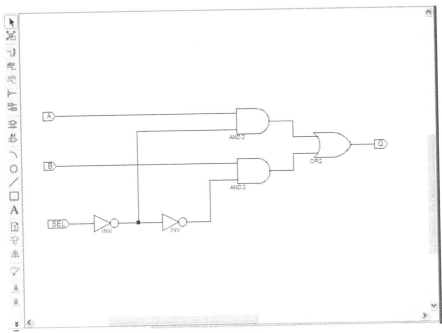

Figure 3-9   *The completed schematic.*

Step 6: Create a User Constraints File

We now come to a step in the process that is specific to the FPGA board hardware that you are using. We are going to use a user constraints file (UCF) to map the NET names in our schematic (SEL, A, B, and Q) to the FPGA pin names such as P123. This mapping is necessary because each of the evaluation boards and plug-in shields will use different FPGA pins.

The UCF is created as a source file. To get back to the original Project View, click the "Design" tab at the bottom of the Design View. Right-click on "data_selector" and again select "New Source..." to open the New Source Wizard. This time select "Implementation Constraints File" (ISE's name for a UCF), and enter the file name "data_selector_elbert," "data_selector_mojo," or "data_selector_papilio" depending on which board you are using. Also set the location to be inside the "src" directory, as shown in Figure 3-10. Click "Next," and finish the Wizard.

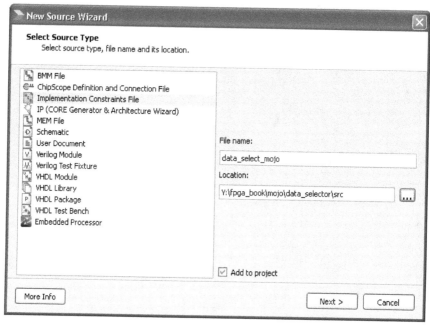

Figure 3-10   *Creating a user constraints file.*

This will open an empty text editor window where you need to type the following text if you are using a Mojo and IO Shield:

```
# User Constraint File for Data Selector on Mojo IO Shield

# DIP Switch 0 is selector for channels A and B
# Left Push Switch is input A
# Right Push Switch is input B
# LED 0 is output

NET "SEL" LOC = "P120" | PULLDOWN;
NET "A" LOC = "P142" | PULLDOWN;
NET "B" LOC = "P138" | PULLDOWN;

NET "Q" LOC = "P97";
```

Don't worry if you have an Elbert or Papilio—they are much the same—you will just need to change the pin numbers and a few other options, as you will see shortly.

The lines that begin with a # are comment lines. That is, like the lines of program code starting with // in C and Java (and Verilog), which take no part in the functioning of the program, the lines starting with # are not part of the configuration information; they are just to make it easier to see what's going on.

After that, there are a number of NET commands that define the links from the net name specified in the schematic to the pin location (LOC) in the actual hardware. Finally, at the end of the lines for the switches, there is a bar symbol followed by the word PULLDOWN to enable the internal pull-down resistor for that input pin.

If you refer to Appendices B, C, and D, you will find all the information you need to write the UCF for your FPGA board. Thus, for example, in Appendix C, which is for the Mojo and Mojo IO Shield, Figure C-2 shows the schematic for the push switches on the Mojo IO board and is repeated here as Figure 3-11. This tells you, for example, that the button labeled "LEFT" on the Mojo IO board is connected to pin P142 of the FPGA.

Here is the very similar UCF for the Elbert 2:

```
# User Constraint File for Data Selector on Elbert 2

# DIP Switch 8 is selector for channels A and B
# Left Push Switch (SW1) is input A
# Right Push Switch (SW2) is input B
```

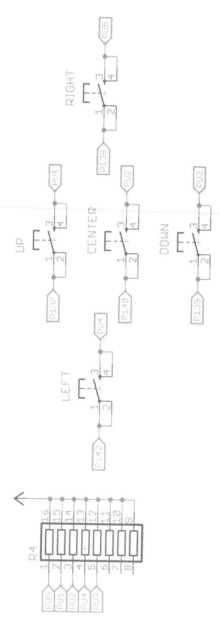

Figure 3-11 *Mojo IO button schematic.*

38

```
# LED D1 is output

NET "SEL" LOC = "P70" | PULLUP;
NET "A" LOC = "P80" | PULLUP;
NET "B" LOC = "P79" | PULLUP;

NET "Q" LOC = "P55";
```

Apart from different FPGA pin allocations, the switch pins are all set to PULLUP rather than PULLDOWN because unlike the Mojo IO Shield, the Elbert switches switch to ground.

Finally, here is the UCF for the Papilio:

```
# User Constraint File for Data Selector on Papilio Logic Mega Wing
# Slide Switch 0 is selector for channels A and B
# Left Push Switch on the joystick input A
# Right Push Switch on the joystick input B
# LED 0 is output

NET "SEL" LOC = "P91";
NET "A" LOC = "P34" | PULLUP;
NET "B" LOC = "P36" | PULLUP;

NET "Q" LOC = "P5";
```

The slide switch for the Papilio LogicStart MegaWing does not require either a pull-up or a pull-down resistor to be enabled; however, the joystick switches require pull-ups (see Appendix D).

Step 7: Generate the .bit File

You are now ready to synthesize the design and generate the programming file to be downloaded onto your board. So select the "data_selector" entry in the hierarchy, and a number of options for things to do will appear below it in the Processes section. One of those processes will be "Generate Programming File" (Figure 3-12). Right-click on this option, and select "Run."

If all is well, lots of text will appear in the Console as the programming file is generated. If there are any errors, this is where they will appear, so read any error messages carefully. They should point to where any problems are located. Often you will find that an error is a misplaced punctuation sign or an incorrectly spelled name in the UCF.

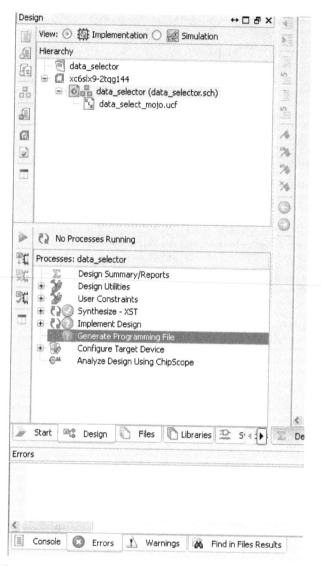

Figure 3-12 *Generating the programming file.*

Step 8: Program Your Board

The end result of all this activity will be a file within the working directory of the project called data_selector.bit. It is this file that we need to transfer onto your board using the loader specific to your board (that you downloaded earlier). The process is the same whatever loader you are using:

- Connect your board to your computer using a USB cable.
- Make sure that the correct serial port is selected in the loader's user interface.
- Select the .bit file to be installed on the FPGA board.
- Install the .bit file on the FPGA board by pressing the "Program, Load or Run" button depending on your board's loader program.

The three loaders' tools are shown in Figure 3-13.

The Mojo loader also offers you the option of "Store to Flash" (select this) and an "Erase" button. You do not need to click the "Erase" button; the "Load" button will erase before loading. One idiosyncrasy with the Mojo loader is that the files selector defaults to .bin rather than .bit, so to select the generated .bit file, you will have to set the filter to "all files."

The Papilio Loader has an option to write to SPI Flash, FPGA, or Disk File; select "SPI Flash."

Testing the Result

Figure 3-14 shows a close-up of the Elbert 2 board with the controls for testing the data selector shown. Slide switch 8 selects either button SW1 or button SW2 to control LED D1.

Put slide switch 8 into the left ON position. Initially, you should see LED D1 on the Elbert V2 light up. If you press SW1, the LED will turn off. Release the button, and LED D1 will turn on again. Pressing SW2 will have no effect. Now put slide switch 8 into the OFF position, and you will notice that SW1 no longer has any effect, but SW2 does alter the LED's state.

This is the data selector working as it should. The logic is a little confusing because the push-switch inputs are effectively inverted as they are pulled to GND when you press the button.

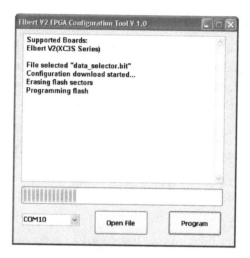

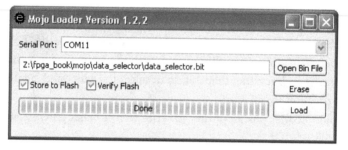

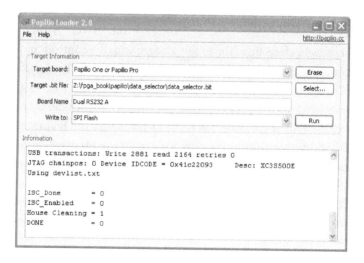

Figure 3-13   *FPGA board loader tools: (top) Elbert; (middle) Mojo; (bottom) Papilio.*

Figure 3-14 *Testing the data selector on an Elbert.*

Figure 3-15 shows the same experiment using the Mojo. This time LED 0 on the IO Shield is used as the output Q. Slide switch 0, selector SEL, and inputs A and B use the left and right push buttons, respectively. Carry out the same test procedure as described for the Elbert 2 to verify that the data selector is working.

Figure 3-15 *Testing the data selector on a Mojo.*

When testing on the Mojo, the push buttons are HIGH when pressed, so the inputs are not inverted.

The Papilio LogicStart MegaWing uses a joystick to provide inputs A and B. Move the joystick to the left for A and to the right for B (Figure 3-16).

A 4-Bit Counter Example

In Chapter 2 we saw how a counter could be built using four D flip-flops. Let's build that counter using the schematics editor in a new project.

Figure 3-16 *Testing the data selector on a Papilio.*

As you did with the "selector" example, start by creating a new project. Give it the name "counter." You should find that when you run the New Project Wizard, this time it remembers all the project settings from the last project, which means that you can accept the default settings and they will be correct for your board. If you would prefer to load up the finished project, then you can find it in the downloads for this book (see Chapter 2) in Project Folder ch03_counter.

Drawing the Schematic

Create a new Schematic source ("counter"), and draw the schematic. It can be hard to find the right symbols, so you will probably need to drop a selection of symbols onto the canvas before you find the right ones. Delete the ones that you don't want by selecting them and then pressing the DELETE key. The symbol we used for each D Flip-flop was found in the category Flip_Flop and called fd. The D flip-flops in ISE's library do not have inverting outputs, so use four NOT (inv) gates to provide the inverted output. Add some IO markers for Clock and QA to QD. The end result should look like Figure 3-17.

Implementation Constraints Files

You also need to create an implementation constraints file (or UCF) and place the following contents in it. Here is the one for the Elbert 2:

```
# User Constraint File for 4 digit counter on Elbert 2

# Left Push Switch (SW1) is Clock
# LEDs D1 to D4 are outputs

NET "Clock" LOC = "P80" | PULLUP | CLOCK_DEDICATED_ROUTE = FALSE;

NET "QA" LOC = "P55";
NET "QB" LOC = "P54";
NET "QC" LOC = "P51";
NET "QD" LOC = "P50";
```

The new first line specifies that Clock is a clock signal that should not be given a dedicated routing. This is so because it is connected to a push button and not to the high-frequency built-in clock. Later in this book you will see that, in practice, you will use synchronous designs that change state only with each tick of the system clock. For now, though, let's just ignore that and make a simple counter.

The UCF for the Mojo is almost identical and just requires the pin names to be changed:

```
# User Constraint File for 4 digit counter on Mojo

# Right Push Switch is Clock
# LEDs 0 to 3 are outputs

NET "Clock" LOC = "P138" | PULLDOWN | CLOCK_DEDICATED_ROUTE = FALSE;

NET "QA" LOC = "P97";
NET "QB" LOC = "P98";
NET "QC" LOC = "P94";
NET "QD" LOC = "P95";
```

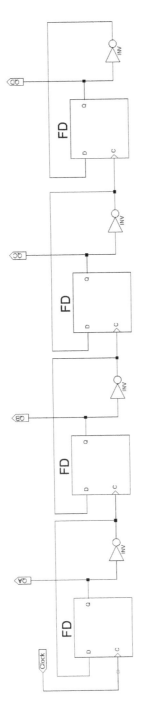

Figure 3-17 Schematic for a 4-bit counter.

The same is true for the UCF for the Papilio:

```
# User Constraint File for 4 digit counter on Papilio and LogicStart
# MegaWing

# Joystick select Push Switch is Clock
# LEDs 0 to 3 are outputs

NET "Clock" LOC = "P22" | PULLUP | CLOCK_DEDICATED_ROUTE = FALSE;

NET "QA" LOC = "P5";
NET "QB" LOC = "P9";
NET "QC" LOC = "P10";
NET "QD" LOC = "P11";
```

Testing the Counter

Generate the programming file, and then deploy it using the loader tool for your board. Note that you will probably see some warnings from ISE, but you can disregard these for now.

Press whatever push button you are using for the Clock signal, and you will see the LEDs count in binary. You will also notice that there is a fair bit of key bouncing from the push switches, and the LEDs may skip past some of the numbers. In Chapter 5 we will show you how to "debounce" such button presses to prevent false triggerings.

Summary

The two example projects in this chapter that use the schematic editor are simple enough to make the schematic editor practical, but as the complexity of your designs grows, it becomes quicker to use Verilog to define your logic. In Chapter 4 we will start by reimplementing the two examples here using Verilog and then move on to look at some more complex examples.

4

Introducing Verilog

Verilog is a hardware description language (HDL). Along with its rival language, VHDL, it is the most common way to program an FPGA. You probably found that programming a FPGA using a schematic is familiar and easy to understand, so why would you want to learn a complicated programming language to do the same thing? Well, the answer is that, actually, as designs become more and more complex, it is easier to represent a design using a programming language than to draw it.

Verilog looks like a programming language, and indeed, you will find "if" statements, code blocks, and other software-like constructions, including the ability to add and subtract numbers.

The examples in this chapter, along with all the examples in this book, can be downloaded from the GitHub repository for this book, as described at the end of Chapter 2.

Modules

Software programmers will recognize a Verilog module as being very like a *class* in object-oriented programming. It defines a collection of logic with public and private properties that can be instantiated a number of times in your design.

If electronics is more your thing, think of a module as a subassembly of the design with defined connections so that it can be wired up to other modules or, if you like, an IC. A simple design may be all contained in a single module, but when things start to get a little complex, the design will become a collection of modules that are then interconnected. In this way, you will also be able to make use of other people's modules in your designs.

49

Wires, Registers, and Buses

What would be variables in a conventional programming language are wires (connecting one thing to another) or registers (which store state and are therefore more like a programming variable) in Verilog. A wire and a register refer to a single binary digit. Often you want to work on more than one bit at a time, so you can group a number of bits into a *vector* and operate on the vector as a whole. This is rather like using a word of arbitrary length in a conventional programming language. When defining such vectors, the upper and lower bits are specified. The following example defines an 8-bit counter:

```
reg[7:0] counter;
```

Parallel Execution

Because Verilog is describing hardware rather than software, there is an implicit parallelism in Verilog. If you have three counters in a design, all connected to different clocks, that's just fine. Each will do its own thing. It is not like using a microcontroller, where there is a single thread of execution.

Number Format

A lot of the time in Verilog, you will be dealing with a vector, and it is convenient to assign values using numbers of any bit size in any radix (number base). To accomplish this, Verilog uses a special number syntax. If you do not specify the number of bits and the radix, then the number is assumed to be decimal, and unused bits are set to 0. The number format starts with the number of bits, then there is an apostrophe, followed by a radix indicator (b = binary; h = hex; and d = decimal), and this is followed by the number constant.

Here are some Verilog integer constants:

4'b1011	four-digit binary constant
8'hF2	8-bit hex constant
8'd123	decimal number 123 represented in 8 bits
123	decimal number 123 with no defined bit size (ISE will do its best to guess the size)

Data Selector in Verilog

Rather than just looking at Verilog in isolation, let's combine it with learning how to use it in ISE. Either follow along with the instructions here to create the project from scratch, or use the project from this book's downloads (see Chapter 2). The Project Folder in the downloads is called ch04_data_selector_verilog.

The first step is to create a new project. This time when the New Project Wizard appears (Figure 4-1), give it the name "data_selector_verilog," and change the drop-down list at the bottom (Top-Level Source Type) to be "HDL" (hardware definition language). This is a change from when you created a new project for a schematic design.

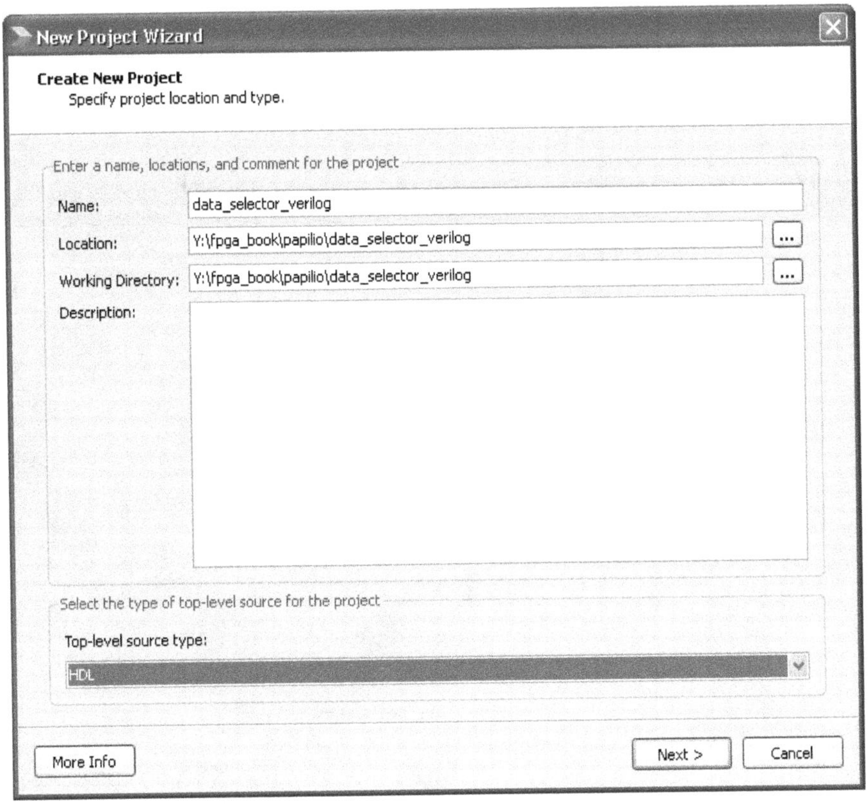

Figure 4-1 *The New Project Wizard.*

When you create the project, remember to change the settings for your board using Appendices B to D. If you are sticking to just one of these boards, then ISE helpfully should remember the settings from the last project you created.

Now we need to create a new source file for the Verilog version of the data selector. So right-click on the project, and select the option "New Source ...." This will open the New Source Wizard (Figure 4-2).

Select a source type of Verilog module and give the source the name "data_selector," as you did with the schematic design; then set the directory to be "src." When you click "Next," you will be prompted to define the inputs and outputs to the module (Figure 4-3). This step generates some code for the Verilog module to get you started. If you prefer, you can just click "Next" without adding any inputs or outputs and type in the Verilog text directly. However, it does save a little typing, so for now, let's use the Wizard.

Use the New Source Wizard window to define three inputs (A, B, and SEL) and one output (Q), as shown in Figure 4-3. Click "Next," and then after the summary screen, click "Finish." The Wizard will then generate a template file for your

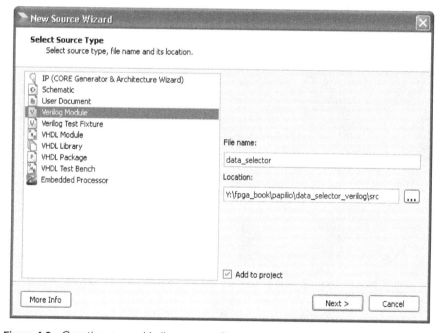

Figure 4-2 *Creating a new Verilog source file.*

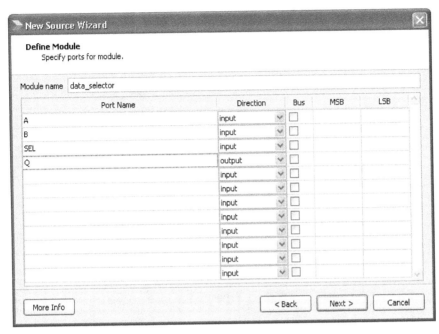

Figure 4-3 *Defining inputs and outputs for the new Verilog source.*

Verilog module using the information you entered (Figure 4-4). This file has the standard extension of .v for Verilog.

At present, this module does not actually do anything, and it has a large block of comment at the top that you might like to delete for clarity. Let's analyze what has been generated by the New Source Wizard. Here is the code that was generated:

```
module data_selector(
  input A,
  input B,
  input SEL,
  output Q
  );

endmodule
```

The module starts with the module keyword and is followed by the name of the module. Inside the parentheses are the inputs and outputs to the module. The word endmodule marks the end of the module definition.

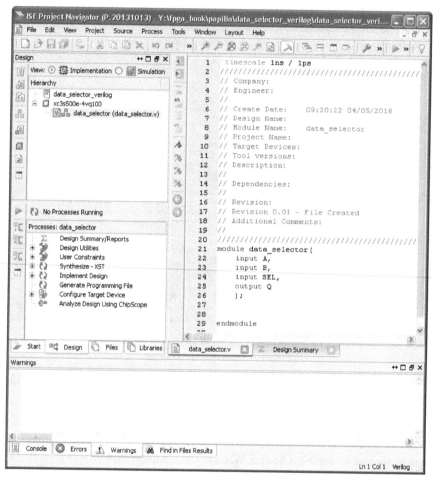

Figure 4-4 *The generated module code.*

Modify the text so that it appears as follows. Note that the additions are marked in boldface.

```
module data_selector(
 input A,
 input B,
 input SEL,
 output reg Q
 );
```

```
always @(A or B or SEL)
begin
  if (SEL)
    Q = A;
  else
    Q = B;
end

endmodule
```

The first change is the addition of the word `reg` to the output definition for Q. This indicates that Q is a register and therefore can be modified.

The other addition is the `always` block. Immediately after `always` is the "sensitivity" list that follows @. This specifies the signals (separated by the word `or`) to which the `always` block is sensitive. That is, the code between `begin` and `end` comes into play. It is tempting to think of this code as if it were a programming language rather than a hardware definition language.

If SEL is 1, then Q will be assigned to whatever the state of A is. Otherwise, Q will be set to the value at input B. This is exactly what the selector should do.

That's all there is to the Verilog for the "data_selector" example; however, you still need an implementation constraints file or UCF if you want to try out the example. The one that you created for the schematic version of this project will work just fine. Copy your .ucf file from the src directory of your schematic data_ selector project into the src directory of your data_selector_verilog project, and then right-click in the ISE hierarchy area, and select "Add Source...." Then navigate to the .ucf file that you just copied into the data_selector_verilog folder.

Build the project, and then install it on your board in the same way you did for the schematic project. The project should work in the same way.

A Counter in Verilog

The "counter schematic" project can also be implemented in Verilog. Either follow along with the instructions here to create the project from scratch, or you can use the project from the book's downloads (see Chapter 2). The project folder in the downloads is called ch04_counter_verilog.

This time, when you create the new project you could call it "counter_verilog." Create a new Verilog module source file in a new src directory (call it "counter"), and add the inputs and outputs as shown in Figure 4-5.

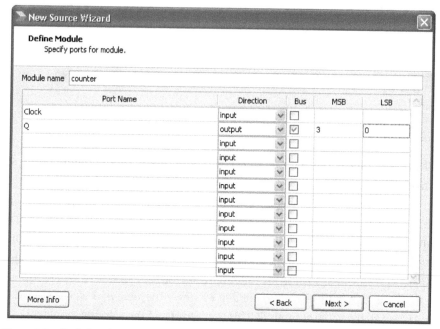

Figure 4-5 *Defining the inputs and outputs for the counter.*

The output Q is defined as being a *bus* by checking the "Bus" checkbox. The "MSB" column indicates its most significant bit number (in this case 3), and the least significant bit (LSB) of 0 is entered in the "LSB" column.

Finish the Wizard, and the generated code will start to look like this:

```
module counter(
 input Clock,
 output [3:0] Q
 );
```

You now need to add the counting logic for the counter, so edit the code to be:

```
module counter(
 input Clock,
 output reg [3:0] Q
 );

always @(posedge Clock)
begin
 Q <= Q + 1;
```

end

```
endmodule
```

The added code is shown in boldface. The sensitivity list in the `always` block includes positive edges of the Clock signal. Whenever the Clock signal goes high, the code between `begin` and `end` comes into play. This simply adds 1 to Q.

When adding 1 to Q the `<=` rather than just `=` operator is used. This type of assignment is called *nonblocking* and should always be used with sequential logic. The reason for this will become apparent later in this chapter when you look at synchronous logic.

Note that in this case, since 0 and 1 are the same in any radix, we have not specified the radix or number of bits in the number constants.

You now need to add a .ucf file for the project that is a modified version of the .ucf file from the schematic-based project that you created in the original counter project in Chapter 3. The only change to this is that now the output is a bus Q[0] to Q[3] rather than separate outputs QA to QD. The version of this file for the Papilio is listed next, with the changes from the schematic version highlighted in boldface.

```
# User Constraint File for 4 digit counter on Papilio and LogicStart
# MegaWing

# Joystick select Push Switch is Clock
# LEDs 0 to 3 are outputs

NET "Clock" LOC = "P22" | PULLUP | CLOCK_DEDICATED_ROUTE = TRUE;

NET "Q[0]" LOC = "P5";
NET "Q[1]" LOC = "P9";
NET "Q[2]" LOC = "P10";
NET "Q[3]" LOC = "P11";
```

This is very similar to the one for the schematic-based counter, but in this case the separate bits of the Q bus are linked to the LEDs using square bracket notation to indicate the bit linked to a particular LED. Rather like arrays in many programming languages, the square bracket notation allows access to individual bits of the bus.

Generate the binary file, and install it on your board. Then you should have something that behaves just like the schematic version.

Synchronous Logic

The preceding counter example illustrates how easy it is to define some hardware in Verilog. However, it is missing an important feature that you will find in almost any Verilog example that you would care to look at. That is, it is not *synchronized* with a clock.

The example works, but only because it is so simple. The problem that arises, as soon as projects become even slightly more complex than this, is that signals take different times to propagate through logic gates, and this means that an output that depends on inputs from many other parts of the system and possibly even the output itself will take time to settle to its final value. It may have glitchy pulses that should be ignored by other parts of the system. Such outputs are described as *metastable*.

This is why you will have seen some warning messages as you built the project, not least because the clock input to the counter appears to ISE as a synchronizing clock input, but we are just using it as a general input connected to a switch.

The solution to this problem of metastable outputs is to use a system-wide clock (usually tens of megahertz). Everything then happens each time the clock ticks. This means that any metastable outputs will have had time to settle before their value is sampled. The <= operator that you used in the counter example in preference to = ensures that all assignments using <= within an `always` block happen at the same time (in the same clock cycle.) This gives time for all the outputs in the system to settle well before the next clock cycle occurs.

Summary

You should now be starting to get familiar with the ISE Design Suite and UCFs. The Verilog examples in this chapter are both implemented as single modules. In Chapter 5 we will look at a more complex example where the design for a multiplexed LED display divides up into a number of modules that are then combined to build the project. This design will also convert what we have done so far into a synchronous design.

5

Modular Verilog

When designing a complex system for a FPGA, there is nothing to stop you from putting all your Verilog code into one module. However, splitting things up makes it easier for others to understand what you have done. This is so because they can assume that the component modules perform the role they are supposed to and therefore see a bigger picture of how all the modules work together before getting into the nitty-gritty of how each one works. Breaking things up into a number of modules also makes it a lot easier to take a module that you used in one project and use it in another or to share it with someone else to use in their project.

When you create a project with more than one module, you will always have a top-level module. This is the module that brings all the submodules together as well as the module that will have a UCF associated with it to map the IO pins of the FPGA to the signals in the design.

In this chapter, you will start by using a seven-segment decoder module along with the counter module of Chapter 4 and expanding the example until eventually you have a simple decimal counter that counts up and down on the seven-segment multidigit display when you press the UP and DOWN switches.

You can either build the projects up yourself following the instructions here or download the completed projects from GitHub, as described at the end of Chapter 2.

A Seven-Segment Decoder

The first reusable module that we are going to build is a seven-segment decoder. This will have a 4-bit input. The number at this input or, rather, the numbers 0 to 9 will be decoded into the correct segment pattern to display the number on a seven-segment display.

Figure 5-1 shows how the segments of a seven-segment display are organized. Each segment is given a letter A–G with an additional connection to the decimal point (DP).

Create a new project called decoder_7_seg. You can also find this project in the downloads for this book where it is called ch05_decoder_7_seg. See the section at the end of Chapter 2 for downloading this book's examples.

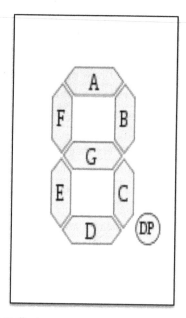

Figure 5-1   *Seven-segment display.*

Create a new Verilog source file in an src directory, and call the file "decoder_7_seg." Write or paste the following code into it:

```verilog
module decoder_7_seg(
    input CLK,
    input [3:0] D,
    output reg [7:0] SEG
    );

always @(posedge CLK)
begin
  case(D)
    4'd0: SEG <= 8'b00000011;
    4'd1: SEG <= 8'b10011111;
    4'd2: SEG <= 8'b00100101;
    4'd3: SEG <= 8'b00001101;
    4'd4: SEG <= 8'b10011001;
    4'd5: SEG <= 8'b01001001;
    4'd6: SEG <= 8'b01000001;
    4'd7: SEG <= 8'b00011111;
    4'd8: SEG <= 8'b00000001;
    4'd9: SEG <= 8'b00001001;
    default: SEG <= 8'b11111111;
  endcase
end

endmodule
```

Starting with the module header, there is a CLK (clock) input because we want the module to synchronize with the FPGA clock. There is also a 4-bit input D that will contain the number to be displayed and an 8-bit register output that will be the bit pattern to appear on the segments of the LED.

The always block synchronizes on the CLK and contains a new construct called a *case statement*. The case statement is a more concise way of writing a whole load of if commands that all have the same value as a condition. Here input D is the value of interest to the case statement. If D is 0, then the output segments (SEG) are set to the bit pattern 00000011. A zero indicates that the segment should be lit. Thus, referring to Figure 5-1, this says that segments a, b, c, d, e, and f should all be lit, leaving just g (in the middle) and the decimal place unlit. Each value of D has its own bit pattern.

The `default` option at the end of the case statement defines what SEG should be set to if none of the values of D match. Because we have a 4-bit value in D, this means that any value over decimal 9 will result in all the segments being off.

This module is designed as a useful component to drop into other projects. To try it out, you need to create another module (one not designed for reuse) that will use this module but also do things like disabling all the digits except one. So create a new source file called seg_test, and put the following code into it:

```
module seg_test(
    input CLK,
    input [3:0] D,
    output [7:0] SEG,
    output [3:0]DIGIT
    );

assign DIGIT = 4'b1110;

decoder_7_seg decoder(.CLK (CLK), .SEG (SEG), .D (D));
endmodule
```

This module's inputs and outputs are

- CLK: clock
- D: Data, 4 bits, each of which will be allocated to a slide switch
- SEG: 8 bits, one for each segment
- DIGITS: 4 bits, to control the digits that are to be enabled

Note how DIGITS is initialized to just enable the first digit using the inverted logic of a zero meaning enabled. Then comes the really interesting line:

```
decoder_7_seg decoder(.CLK (CLK), .SEG (SEG), .D (D));
```

This creates a decoder_7_seg called "decoder," and its parameters between "(" and ")" specify how "decoder" will be connected to the wires and registers in the seg_test module. The names starting with a dot name the parameter in the decoder and then inside a further set of parentheses are references to "wires" or "regs" in seg_test to which the names should be linked.

Finally, you need to create a UCF for the project. The one for the Mojo is listed next; you will find equivalents for the other boards in this book's downloads (see Chapter 2).

```
# User Constraint File for 7-seg decoder implementation on Mojo
# with IO Shield

NET "CLK" LOC = P56;

# 7-segments
NET "SEG[7]" LOC = P5;
NET "SEG[6]" LOC = P8;
NET "SEG[5]" LOC = P144;
NET "SEG[4]" LOC = P143;
NET "SEG[3]" LOC = P2;
NET "SEG[2]" LOC = P6;
NET "SEG[1]" LOC = P1;
NET "SEG[0]" LOC = P141;


# Digits
NET "DIGIT[3]" LOC = P12;
NET "DIGIT[2]" LOC = P7;
NET "DIGIT[1]" LOC = P10;
NET "DIGIT[0]" LOC = P9;


# Inputs to slide switches 0 to 3
NET "D[0]" LOC = P120 | PULLDOWN;
NET "D[1]" LOC = P121 | PULLDOWN;
NET "D[2]" LOC = P118 | PULLDOWN;
NET "D[3]" LOC = P119 | PULLDOWN;
```

To test out the project, upload it onto your board, and then try dialing in different values on the slide switches. Figure 5-2 shows this in action.

Button Debouncing

When you were experimenting with the counter project, you may have noticed some switch bounce that caused the counter to skip over some numbers. We can do something about this by creating a debouncing module that we can use in any project where we use a switch.

When it comes to push switches, there are actually three problems to solve. First, there is a problem that a button press is by its nature not going to be synchronized to the clock. Second, we have the problem of actually removing any

Figure 5-2   *Testing the seven-segment decoder module.*

unwanted transitions resulting from the mechanical contacts of the switch bouncing. Finally, there is the problem that you are usually interested in triggering things on a transition for a button from OFF to ON (or vice versa) rather than at any particular time it is ON or OFF (its state).

We can lump these three features together into one very useful module that takes a potentially bouncy switch as an input and gives us as an output a clean, synchronized binary state of the button, along with a pair of useful extra outputs that go high for a single clock cycle when the button transitions. You can find the project illustrating switch debouncing in the project "debounce." You may wish to have this open in ISE because the code is explained.

The code for this is adapted from www.fpga4fun.com/Debouncer.html and www.eecs.umich.edu/courses/eecs270/270lab/270_docs/debounce.html.

The first step is to synchronize the push switch to the clock using a pair of registers. Here is the listing for debouncer.v together with explanations of the code. You can find a project illustrating the use of this module in the download ch05_debouncer file.

```
module debouncer(
    input CLK,
    input switch_input,
```

```
   output reg state,
   output trans_up,
   output trans_dn
   );

// Synchronize the switch input to the clock
reg sync_0, sync_1;
always @(posedge CLK)
begin
  sync_0 = switch_input;
end

always @(posedge CLK)
begin
  sync_1 = sync_0;
end
```

The output of the second register (sync_1) is then used for the debouncing code:

```
// Debounce the switch
reg [16:0] count;
wire idle = (state == sync_1);
wire finished = &count;           // true when all bits of count are 1's
```

The debouncing itself works by using a timer (in this case a 16-bit timer). This then ignores any transitions in the switch signal until the timer has counted to its maximum value. In the case of a 16-bit timer, this is 65,536 clock cycles, which for the 50-MHz clock of a Mojo is 65,536/50,000,000 = 1.3 ms. On an Elbert 2 with its 12-MHz clock, this will be 5.4 ms, and on the 32-MHz Papilio One, it will be 2 ms. Generally, the switch bouncing will settle out in well under a millisecond, and 5.4 ms is still a very short time as far as a user is concerned, so the code does not really need to be modified for the other boards used in this book despite their different clock speeds.

Sizing Counter Registers

Most FPGA projects will require one or more counters; in a conventional programming language, integer variables are of a fixed size (say 32 or 64 bits), but in the world of FPGAs, every vector can be of a different size. Deciding how many bits you need requires a little math. A 1-bit counter can only hold a maximum value of 0 to 1, a 2-bit counter has a

range of 0 to 3, a 3-bit counter 0 to 8. The upper limit of a counter value
is one less than 2 raised to the power of N, where N is the number of
bits in the counter. Table 5-1 will help you to select the number of bits
you need to hold a particular value.

Number of Bits	Maximum Value (Decimal)
1	1
2	3
3	7
4	15
5	31
6	63
7	127
8	255
9	511
10	1,023
11	2,047
12	4,095
13	8,191
14	16,383
15	32,767
16	65,535
17	131,071
18	262,143
19	524,287
20	1,048,575

Table 5-1   *Sizing Counters*

The idle wire's value is set to be the result of comparing state with sync_1. If
they are the same, then no switch transition in is progress, and idle will have the
value 1; otherwise, it will have the value 0. In common with such programming
languages as Java and C, the == operator is used to compare values to see if they
are equal.

To detect that the counter is maxed out, the logical & operator is used. When
used on a multibit register such as "count," the result is all the bits being ANDed
together. This state is associated with the wire finished.

```
always @(posedge CLK)
begin
  if (idle)
  begin
    count <= 0;
  end
  else
  begin
    count <= count + 1;
    if (finished)
    begin
      state <= ~state;
    end
  end
end
```

The main `always` block is synced to the positive edge of the clock and, if nothing is happening, resets count to 0. If a transition is in progress, then the counter is incremented. When the counter is finished, the `state` output is toggled. The ~ symbol means invert in Verilog.

It just remains to define the useful extra output values of `trans_dn` and `trans_up`, which will become high for one clock cycle as the button is pressed or released, respectively. This is accomplished by ANDing together the inverted output of the `idle` signal, whether the timer has "finished" and then `state` is output or, in the case of `trans_dn`, the inverse of `state`.

```
assign trans_dn = ~idle & finished & ~state;
assign trans_up = ~idle & finished & state;

endmodule
```

The test program for the debounce code illustrates how the module can be used for all three possible outputs. It uses two push buttons, each of which toggles the state of separate LEDs, one on `trans_dn` (button pressed) and one on `trans_up` (buttton released). A third LED simply mirrors the state of the first button.

```
module debounce(
    input CLK,
    input switch_a,
    input switch_b,
    output reg led_a,
```

```
output reg led_b,
output reg led_c
);
```

Three wires are defined, s_a_dn (switch a down), s_b_up and s_a_state. These wires are then linked to three debouncers d1 to d3. Looking at the first of these debouncer lines of code, you can see that the debouncer has been given the name d1, and then there are a number of parameters enclosed in parentheses. Each parameter is in the form of a pair, and each of these pairs is separated by a comma. The first parameter, .CLK (CLK), links the CLK signal of the debouncer (the CLK preceded by a period) to the CLK used in this debounce test module.

The second parameter, .switch_input (switch_a), links the switch_input of the debouncer to switch_a. The final parameter, .trans_dn (s_a_dn), connects the trans_dn output of the debouncer to the wire s_a_dn. This repeats for the other two debouncer instances. However, these instances use the trans_up and state outputs of the debouncers, linking them to the wires s_b_up and s_a_state, respectively.

```
wire s_a_dn, s_b_up, s_a_state;
debouncer d1(.CLK (CLK), .switch_input (switch_a),
            .trans_dn (s_a_dn));
debouncer d2(.CLK (CLK), .switch_input (switch_b),
            .trans_up (s_b_up));
debouncer d3(.CLK (CLK), .switch_input (switch_a),
            .state (s_a_state));
```

The always block is synced to the positive edge of the clock and uses the wires linked to the outputs of the three debouncers to set the three LEDs.

```
always @(posedge CLK)
begin
  if (s_a_dn)
  begin
    led_a <= ~ led_a;
  end
  if (s_b_up)
  begin
    led_b <= ~ led_b;
  end
  led_c <= s_a_state;
```

```
end

endmodule
```

Now that you have tamed our switch inputs, you do not need to worry about how this module works; you can just use it within any of your projects that needs switch debouncing. To illustrate this, in the next section, you will bring together both the debouncer and decoder_7_seg modules to make a third module that is a four-digit (three in the case of the Elbert 2) counter display.

Multiplexed Seven-Segment Display and Counter

This example (Figure 5-3) will display a number on the multidigit seven-segment display. Pressing the "Up" button will increment this number and the "Down" button will reset it to zero. You can find the files for this in the project "ch05_counter_7_seg."

Figure 5-3 *A multiplexed seven-segment display.*

All three FPGA boards have *multiplexed* seven-segment LED displays. In the case of the Mojo (with IO Shield) and the Papilio (with LogicStart MegaWing), these are four-digit displays, and the Elbert 2 is a three-digit display. In all cases, the displays are wired up in much the same way. Figure 5-4 shows the schematic for the Mojo IO Shield's display; the other boards are much the same but use different pins.

Each segment of a particular digit is connected to the same segments on the other digits. In other words, all the segments A are connected together, all the segments B are connected together, and so on. Each segment is then connected via a resistor (shown as R5 in Figure 5-4) that is in turn connected to a general-purpose input-output (GPIO) pin.

This means that if you say set GPIO pin P8 HIGH, then all segments B would be enabled. Clearly, you need to be able to display different numbers on each digit. To do this, each digit is enabled in turn using the separate digit control pins and the segment pattern for that digit set before moving onto the next digit and setting a different segment pattern. This happens so fast that you don't see any flicker—it just appears that each digit is displaying a different segment pattern.

Project Structure

Before wading into the detail, it's worth taking a step back and looking at how the various modules used in this project will relate to each other (Figure 5-5).

The module counter_7_seg is called the *top-level module* (more on this later). It is to this module that the UCF for your board will be attached. This module implements the logic of the counter: incrementing its value, and resetting it and making use of other modules that operate the display.

The counter_7_seg module has one instance of `display_7_seg` and two instances of `debouncer` within it. The display_7_seg module will be responsible for multiplexing the display and refreshing it so that it can display all the digits. The module display_7_seg itself contains the decoder_7_seg module that we created earlier. It will use decoder_7_seg to set the segment pattern of each digit in turn.

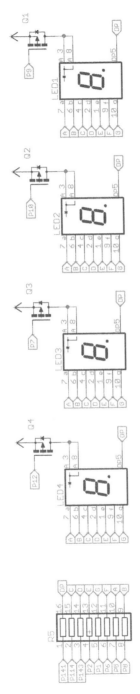

Figure 5-4 *Schematic of the Mojo IO Shield's display.*

71

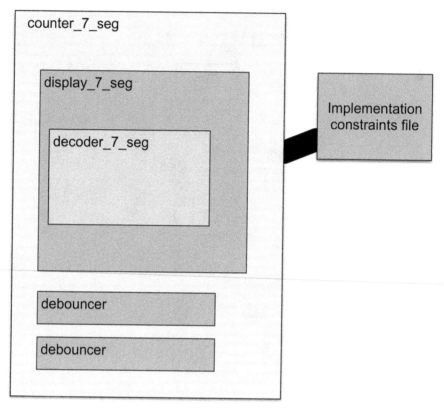

Figure 5-5 *The modules used in the "counter_7_seg" project.*

Display_7_seg

You have already learned about the two modules decoder_7_seg and debouncer, so we can start with the module display_7_seg. Here is the listing for display_7_seg.v together with explanations of the code:

```
module display_7_seg(
    input CLK,
    input [3:0] units, tens, hundreds, thousands,
    output [7:0] SEG,
    output reg [3:0] DIGIT
    );
```

As well as a CLK signal, the module has four (three for the Ebert 2) inputs for each of the digit positions called *units, tens, hundreds,* and *thousands.* The two outputs are for the GPIO pins that will be assigned to control the segments and the digit enable lines.

Three vectors are needed:

- `Digit_data` is a 4-bit register that will be assigned to the 4-bit number to be decoded for the digit currently being processed.

- `Digit_position` is a 2-bit counter that is used to keep track of which digit is currently being displayed.

- `Prescaler` is a 24-bit counter that is used to divide the CLK input to set the refresh rate. The prescaler does not need to be nearly as big as 24 bits, but this gives us options should we use a really fast FPGA.

```
reg [3:0] digit_data;
reg [1:0] digit_posn;
reg [23:0] prescaler;
```

Next, the display_7_seg module needs an instance of decoder_7_seg to convert a 4-bit number into an 8-bit segment pattern. We only need one of these because it will be used for each of the digits in turn.

The D input of the decoder is connected to digit_data, and the SEG vector is passed along again to the decoder:

```
decoder_7_seg decoder(.CLK (CLK), .SEG (SEG), .D (digit_data));
```

Now comes the always block, synched to the clock. This adds 1 to the prescaler counter. If the value of prescaler has reached 50,000, then it's time to refresh the next digit. The 50-MHz clock speed of the Mojo means that this will happen every millisecond. Four digits therefore will take 4 ms, resulting in a refresh frequency of 250 Hz, which is still far too fast for the human eye to see any flicker. You can, if you really want, change this value to suit the other board's clock frequencies, but even with the 12-MHz clock of the Elbert 2, the refresh will be too fast for the flicker to be visible.

```
always @(posedge CLK)
begin
   prescaler <= prescaler + 24'd1;
   if (prescaler == 24'd50000) // 1 kHz
   begin
```

If the prescaler has reached its set value and its time for the next digit refresh, a series of updates now take place, to switch in the appropriate digit data.

First, the prescaler is reset to 0 to start the timer for the next digit refresh, and then digit_posn is incremented. If digit_posn is 0, then digit_data is set to be units. The digit control pins (DIGIT) are then set to enable just the first digit. In this case, the binary pattern 1110 is used. It is 1110 rather than 0001 because the digit control pins are active LOW. A 0 means that that digit is enabled.

The same pattern is repeated for each of the other digit positions:

```
      prescaler <= 0;
      digit_posn <= digit_posn + 2'd1;
      if (digit_posn == 0)
      begin
        digit_data <= units;
        DIGIT <= 4'b1110;
      end
      if (digit_posn == 2'd1)
      begin
        digit_data <= tens;
        DIGIT <= 4'b1101;
      end
      if (digit_posn == 2'd2)
      begin
        digit_data <= hundreds;
        DIGIT <= 4'b1011;
      end
      if (digit_posn == 2'd3)
      begin
        digit_data <= thousands;
        DIGIT <= 4'b0111;
      end
    end
  end
endmodule
```

The 2-bit digit_posn counter will automatically wrap around from 3 to 0, and so does not need to be explicitly reset. The Elbert 2 has only three digits, so the digit_posn counter will need to be reset to 0 after the third digit has been displayed.

Counter_7_seg

The counter_7_seg module is the top-level module that pulls everything together. The module's parameters are all concerned with GPIO pins and will map onto the nets defined in the UCF file. Here is the listing for counter_7_seg.v together with explanations of the code:

```
module counter_7_seg(
    input CLK,
    input switch_up,
    input switch_clear,
    output [7:0] SEG,
    output [3:0] DIGIT
    );
```

The project requires two push switches, one to increment the count and one to clear the display back to 0000. These switches are linked to wires and debounced using debouncer modules.

```
wire s_up, s_clear;
debouncer d1(.CLK (CLK), .switch_input (switch_up), .trans_dn (s_up));
debouncer d2(.CLK (CLK), .switch_input (switch_clear),
                .trans_dn (s_clear));
```

Four 4-bit registers are needed, one for each of the four digits:

```
reg [3:0] units, tens, hundreds, thousands;
```

The display_7_seg instance is wired up to the CLK and the four-digit registers, as well as the SEG and DIGIT connections that will connect to the GPIO pins of the FPGA that control the segments and digit selection:

```
display_7_seg display(.CLK (CLK),
                    .units (units), .tens (tens),
                    .hundreds (hundreds), .thousands (thousands),
                    .SEG (SEG), .DIGIT (DIGIT));
```

Most of the action takes place in the always block. This checks to see if the debounced switch s_up has been pressed. If it has, it adds one to units. There then follows a sequence of ever deeper if statements that check for the overflow of one digit and then increment the next digit and reset the current digit to 0 if this occurs:

```
always @(posedge CLK)
begin
  if (s_up)
  begin
    units <= units + 1;
```

```
    if (units == 9)
    begin
      units <= 0;
      tens <= tens + 1;
      if (tens == 9)
      begin
        tens <= 0;
        hundreds <= hundreds + 1;
        if (hundreds == 9)
        begin
          hundreds <= 0;
          thousands <= thousands + 1;
        end
      end
    end
  end
```

If the "Clear" switch is pressed, then all four digits are zeroed:

```
  if (s_clear)
  begin
    units <= 0;
    tens <= 0;
    hundreds <= 0;
    thousands <= 0;
  end
end

endmodule
```

Specifying Bit Size and Radix

The sharp-eyed among you will notice that I have been a little inconsistent with my number formats when assigning values of 0 or 1. Some would say that you should always specify the number of bits and the number base. In fact, it is good practice to do so, but for a small example such as this, it is not essential. ISE will automatically adjust the bit size of the constant to fit the vector to which it is being assigned, and the numbers 1 and 0 are the same in every number base.

For numbers other than 0 and 1, it is a good idea to select the bit size and radix to avoid any possible confusion. However, there is nothing at all wrong with using decimal—pick whatever number base makes the code easiest to understand.

User Constraints File

You will also need a UCF file. The one for Mojo is listed next. You will find suitable equivalents for the other boards in the downloads for this book.

```
# User Constraint File for 7-seg counter on Mojo with IO Shield


NET "CLK" LOC = P56;


# Switches
NET "switch_up" LOC = "P137" | PULLDOWN;
NET "switch_clear" LOC = "P139" | PULLDOWN;


# 7-segments
NET "SEG[7]" LOC = P5;
NET "SEG[6]" LOC = P8;
NET "SEG[5]" LOC = P144;
NET "SEG[4]" LOC = P143;
NET "SEG[3]" LOC = P2;
NET "SEG[2]" LOC = P6;
NET "SEG[1]" LOC = P1;
NET "SEG[0]" LOC = P141;


# Digits
NET "DIGIT[3]" LOC = P12;
NET "DIGIT[2]" LOC = P7;
NET "DIGIT[1]" LOC = P10;
NET "DIGIT[0]" LOC = P9;
```

Importing Source Code for Modules

There are several ways of getting source code for modules that you have created into a project. ISE occasionally gets confused about which files belong to it, and I have found the least problematic procedure for copying a module from another project (say debounce or decoder_7_seg in this example) is as follows:

1. Create the new project.

2. From your operating system's file explorer, create a "src" directory inside the project directory.

3. Add copies of any module files that this project needs (debouncer.v and decoder_7_seg.v) into this "src" directory.

4. From ISE, use the option "Add Source…" and navigate to each of the .v files you just added to the "src" directory.

Setting the Top-Level Module

The "Design" tab in ISE shows you the relationship between the modules used in your project (Figure 5-6).

You can see in this hierarchical view that everything is correct, with the counter_7_seg module containing two debouncer modules and a display_7_seg module that itself contains a decoder_7_seg module. If you look closely, you can see that the counter_7_seg module has an icon next to it comprising three little squares in a triangular arrangement. This indicates that the counter_7_seg module is the top-level module of the project.

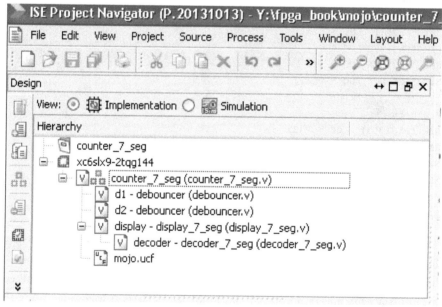

Figure 5-6   *The modules used in the "counter_7_seg" project.*

ISE will often work out for itself what the top-level module is, but sometimes it doesn't, in which case you can specify that a particular module is the top-level module by right-clicking on it and specifying the option "Set As Top Module."

The Three-Digit Version

The Elbert 2 has only three digits rather than the four of the Mojo, so in most cases the differences in the Elbert version are simply reducing the number of digits from four to three. One effect of using three rather than four digits is that in `display_7_seg` the `digit_posn` counter needs to be manually reset to 0 after it reaches the hundreds digit:

```
if (digit_posn == 2'd2)
begin
  digit_data <= hundreds;
  DIGIT <= 4'b1011;
  digit_posn <= 0;
end
```

Testing

Build the project and upload it onto your board. You will find that the number on the display will increment each time you press "Up" and reset to zero when you press the "Clear" button. You will need a lot of patience to test all four digits or even all three on the Elbert 2.

Summary

In this chapter, the potential of the FPGA is starting to reveal itself. In Chapter 6, we will reuse the modules described in this chapter and go on to produce a realistic example of a countdown timer. The timer example also introduces a key concept in the design of FPGA systems called *state machines*.

6

Timer Example

In this chapter, you will build on the general-purpose display_7_seg, decoder_7_seg, and debouncer modules to make a countdown timer that actually functions like a real product. Even an apparently simple device like a timer can be tricky to design. To simplify the design, a representation called *state machines* is often used. This technique is not unique to FPGAs and is a great way of modeling in diagrammatic form just how a system behaves. This chapter also jumps ahead slightly, borrowing a sound-generation module from Chapter 8 to drive a buzzer that will sound once the countdown has finished.

You can find all the code for this project in the downloads for this book (see Chapter 2), and you will find it useful to have the project loaded up in ISE so that you can see the code and, if you feel like it, experiment by modifying it. The project is called "ch06_countdown_timer."

State Machines

State machines have their roots in mathematics and, therefore, like many things in math, sound really clever but are actually pretty simple when it comes down to it. You will also find state machines and drawings of state machines being called *finite state machines* (FSMs) and *state transition diagrams* (somewhat unfortunately STDs).

The basic concept is that a system stays in a stable state until some trigger causes it to *transition* to another state. This transition might cause some actions to occur during the transition. States are drawn on a diagram as boxes or bubbles and transitions as lines with arrows that go either from one box to another or from one box back to itself. The transition lines are drawn with a two-part label on them. The top part of the label is the *condition* for the transition to occur (say, a button being pressed), and then there is a horizontal line, and under the line appear any actions that need to take place during the transition (say, turn a LED on).

There are, of course, other ways of drawing state machines, but this is probably the most common. Figure 6-1 shows the state machine for a fictional coffee maker. It is assumed that the coffee maker has two heating elements, a boiling element and a warming element, with the warming element keeping a hotplate and the coffee warm.

The dot and first transition line indicate the starting state for the machine when it is first turned on, in this case the "Idle" state. The coffee maker will stay in this state until the "Start" button is pressed, at which point the boiling and warming elements will be turned on, and the machine will be in the "Brewing" state until the water reservoir is empty, at which point the power to the heating element will be turned off and the state of the machine will be in the "Keeping Warm" state. At any point, the user can press the "Cancel" button, which will turn off the heating elements and return the machine to its "Idle" state.

This example is very straightforward, and by following the transitions around the diagram, you can get a good feel for how the system will operate in practice. You could also think up other scenarios that could be dealt with—perhaps an "Error" state that is entered if the water reservoir is empty before "Start" is pressed. The condition for such a transition might be "Start pressed AND water reservoir empty."

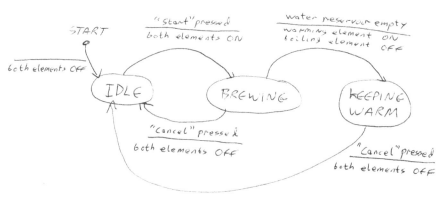

Figure 6-1   *The state machine for a coffee maker.*

Drawing State Machines

I find that hand drawing a state machine on the traditional "back of an envelope" or even a whiteboard is easier than using software, although if you find that you might be returning to the design in the future, you may want to scan or photograph the state machine and save it in the project folder for future reference.

State Machine Design

To allow this example to work on all three example boards, it will use three seven-segment digits (the Elbert 2 has only three digits). Two digits display seconds, and one displays minutes. The "Up/Down" push buttons will set the minutes, a "Start/Stop" button will begin countdown timing, and a "Cancel" button will reset the timer to the number of minutes last used. A buzzer will be wired up to the board to make a noise when the countdown reaches zero. The state machine for this project is shown in Figure 6-2.

The initial state is SETTING, and while in this state, the "Up" and "Down" buttons will add or take away 1 from the digit being displayed in the minutes digit. Once the "Start/Stop" button is pressed, the project moves to the RUNNING state. Pressing "Start/Stop" will immediately take you back to the SETTING state, as will pressing "Cancel." However, pressing "Cancel" will also reset the time. While in

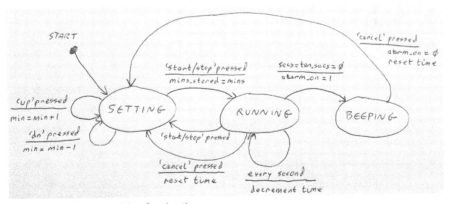

Figure 6-2 *State machine for the timer.*

the SETTING state, every second, the time displayed will be decreased by 1 second until the countdown reaches 000, at which point the BEEPING state is entered.

The BEEPING state sounds the buzzer until the "Cancel" button is pressed, at which point the system returns to the SETTING state, and the time is reset, ready to use the timer again.

Hardware

This project requires an external piezoelectric buzzer that is connected between a GPIO pin and GND. In the case of the Elbert 2 and Mojo, this is connected using male-to-female jumper wires. However, the Papilio fitted with a LogicStart MegaWing makes the GPIO pins inaccessible, so for this board the audio jack is used to generate the tone, and you will need to attach an external amplifier to the audio jack to hear the tone.

You Will Need

In addition to your FPGA board and any Shield/Wing, you are going to need the following items to build this example project. However, if you are prepared to forego the buzzing feature, all you need is your board and shield.

Part	Source
Passive piezoelectric sounder	Adafruit, product ID 160
2× Female-to-male jumper leads	Adafruit, product ID 1953

Construction

The sounder leads may be a little small to be held tightly by the jumper wires, but putting a little bend in them with pliers will ensure a tight fit. Figure 6-3 shows the buzzer attached to the Mojo IO shield, and Figure 6-4 shows it attached to the Elbert 2.

Using a Mojo, the two leads to the buzzer should be connected to GND and P97, situated on the bottom GPIO connector. It does not matter which way around the sounder is connected. To make the project portable, you can add a USB battery backup, as shown in Figure 6-3. When connecting the buzzer to the Elbert 2, one buzzer lead goes to GND on header P1 and the other goes to pin 1 of header P1 (GPIO pin P31).

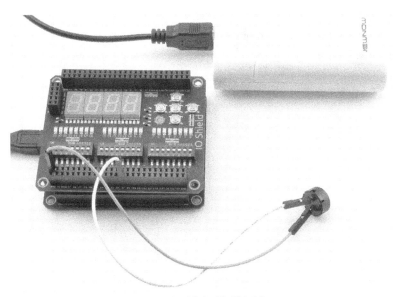

Figure 6-3 *Attaching the buzzer to the Mojo IO Shield.*

Figure 6-4 *Attaching the buzzer to the Elbert 2.*

When the LogicStart MegaWing is fitted onto the Papilio One, you cannot get at the GPIO outputs, so on this board the buzzer tone is played from the audio output jack. You will hear it if you plug the lead from a powered speaker into the board.

Modules

The structure of the modules (Figure 6-5) is fairly similar to that of the "counter_7_seg" project in Chapter 5. However, it does need a few more debouncers for the extra push switches used and also uses a module called "alarm" that will be explained fully in Chapter 8.

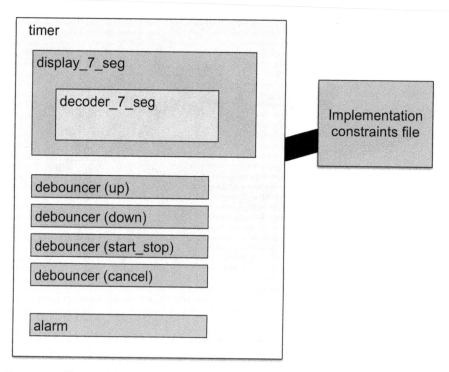

Figure 6-5 *The module structure for the timer project.*

User Constraints File

The user constraints file (UCF) for Mojo and IO Shield is listed next. You will find the files for the Elbert 2 and the Papilio One in the parallel projects for those platforms. The main differences involve just the pin allocations and pullup/pulldown differences to the switch inputs.

```
# User Constraint File for 7-seg counter on Mojo with IO Shield


NET "CLK" LOC = P56;


# Switches
NET "switch_up" LOC = "P137" | PULLDOWN;
NET "switch_dn" LOC = "P139" | PULLDOWN;
NET "switch_cancel" LOC = "P138" | PULLDOWN;
NET "switch_start_stop" LOC = "P140" | PULLDOWN;


# 7-segments
NET "SEG[7]" LOC = P5;
NET "SEG[6]" LOC = P8;
NET "SEG[5]" LOC = P144;
NET "SEG[4]" LOC = P143;
NET "SEG[3]" LOC = P2;
NET "SEG[2]" LOC = P6;
NET "SEG[1]" LOC = P1;
NET "SEG[0]" LOC = P141;


# Digits
NET "DIGIT[3]" LOC = P12;
NET "DIGIT[2]" LOC = P7;
NET "DIGIT[1]" LOC = P10;
NET "DIGIT[0]" LOC = P9;


# Output
NET "BUZZER" LOC = P97; # also LED1
```

Even though the project only uses three digits, all four digits are defined in the UCF so that the unused digit can be blanked out.

A new net BUZZER is defined to generate the tone for a piezoelectric buzzer attached to GPIO pin P97. This pin is, on the Mojo IO Shield, also connected to LED1.

The Timer Module

The top-level timer module brings together all the modules that we have already met in Chapter 5 (debouncer, display_7_seg, etc.) as well as the alarm module that we are borrowing from Chapter 8.

Inputs and Outputs

The declaration for the timer module links the inputs and outputs with the nets in the UCF:

```
module timer(
    input CLK,
    input switch_up,
    input switch_dn,
    input switch_cancel,
    input switch_start_stop,
    output [7:0] SEG,
    output [3:0] DIGIT,
    output BUZZER
    );
```

Push Buttons

The push-button switches are defined in much the same way as in the "counter_7_seg" example from Chapter 5. First, wires are defined for each button, and then a debouncer is allocated to each wire that is attached to the trans_dn (transition down) output of the debouncer:

```
wire s_up, s_dn, s_cancel, s_start_stop;
debouncer d1(.CLK (CLK), .switch_input (switch_up), .trans_dn (s_up));
debouncer d2(.CLK (CLK), .switch_input (switch_dn), .trans_dn (s_dn));
debouncer d3(.CLK (CLK), .switch_input (switch_cancel),
            .trans_dn (s_cancel));
debouncer d4(.CLK (CLK), .switch_input (switch_start_stop),
            .trans_dn (s_start_stop));
```

Alarm Module Instance

The great thing about modular design is that you can make use of a module without having to worry how it works. You just need to know how to use it. So, for now, let's just use the alarm module until you meet it again in Chapter 8.

A register (alarm_on) is used to switch the buzzer on and off. This is linked to the enable input of the alarm instance. This instance is also provided with the CLK input and the GPIO pin to turn the buzzer on:

```
reg alarm_on = 0;
alarm a(.CLK (CLK), .BUZZER (BUZZER), .enable (alarm_on));
```

Modeling Time and the Display

To keep track of the time, separate 4-bit registers are defined for seconds (secs), tens of seconds (ten_secs) and minutes (mins). An additional register (mins_stored) is used to record a copy of the number of minutes set initially, before the timer starts, so that when the timer goes back to its SETTING state, the last time used is remembered.

In the case of the Mojo and the Papilio, the unused digit of the display needs to be blanked out. This is done by setting the value of this digit to 10. The display_7_seg and decoder_7_seg do not display digits over 9, so this will keep this digit blanked out.

The 26-bit register prescaler is used to count the CLK ticks when needed to derive a 1-second tick by counting each FPGA clock cycle until a certain value is reached (see the section "Tasks" later):

```
reg [3:0] secs = 0;
reg [3:0] ten_secs = 0;
reg [3:0] mins = 1;
reg [3:0] mins_stored;
reg [3:0] unused_digit = 4'd10; // digits above 9 not displayed
reg [25:0] prescaler = 0;
```

The display_7_seg module is linked up to the registers described earlier in the appropriate digit positions. Once they are linked like this, we just need to change the register values secs, ten_secs, and mins, and the display will refresh itself:

```
display_7_seg display(.CLK (CLK),
            .units (secs), .tens (ten_secs), .hundreds (mins),
            .thousands (unused_digit),
            .SEG (SEG), .DIGIT (DIGIT));
```

State Machine Implementation

The Verilog code for the state machine follows the state machine diagram of Figure 6-2 very closely. A 2-bit register (state) is used to keep track of the current state. So that you don't have to refer to the states using a number, the localparam command allows you to define three values like constants in conventional programming language called SETTING, RUNNING, and BEEPING to correspond to the three states of the project:

```
// States
localparam SETTING = 0, RUNNING = 1, BEEPING = 2;
reg [1:0] state = SETTING;
```

The key to writing a nice, clean state machine implementation in Verilog is to keep the always block as short as possible by using tasks to move the actions that occur during a transition out of the body of the always block; otherwise, it just gets enormous.

A case statement is used to separate the code for each state. The case statement is an alternative to using lots of if commands if you have a number of conditions that depend on the same value (in this case state). Looking at the first clause in the case statement, we have

```
always @ (posedge CLK)
begin
  case (state)
    SETTING : begin
      handle_settings();
      if (s_start_stop)
      begin
        mins_stored <= mins;
        state <= RUNNING;
      end
    end
```

This performs some logic contained in the task handle_settings that you will see shortly; it then handles the only possible transition out of the SETTING state into the RUNNING state that will occur if the "Start/Stop" button is pressed.

This first action stores away mins into mins_stored before setting the new value of state. The code for the RUNNING state is more complicated because it has three transitions out of this state to handle. It starts by invoking a task decrement_time that will count down by 1 second, and then it handles the other transitions. If the "Start/Stop" button is pressed, it jumps straight back to the SETTING state. The same happens if the "Cancel" button is pressed, but the reset_time task is invoked first.

Finally, if secs, ten_secs, and mins are all zero, the countdown is complete, and the alarm module is enabled to start the buzzer sounding, and the state is set to BEEPING:

```
RUNNING : begin
  decrement_time();
  if (s_start_stop)
  begin
      state <= SETTING;
  end
  if (s_cancel)
  begin
    reset_time();
    state <= SETTING;
  end
  if ((secs == 0) & (ten_secs == 0) & (mins == 0))
  begin
    alarm_on <= 1;
    state <= BEEPING;
  end
end
```

The BEEPING state only has to deal with the transition that occurs if the "Cancel" button is pressed:

```
    BEEPING : begin
      if (s_cancel)
      begin
        alarm_on <= 0;
        state <= SETTING;
        reset_time();
      end
    end
  endcase
end
```

Tasks

The remainder of the code in timer.v implements the tasks that are used in the always block. In Verilog, you use a task or its relative a function for much the same reasons that you would use a function in a programming language. They allow you to structure your code and make it more readable by separating out features into more manageable chunks. A *function* is like a *task* except that it returns a value (i.e., has an output).

You may have noticed that the reset_time task is invoked from more than one place in the always block. By separating it into a task of its own, not only have we made the code easier to understand, but we have also eliminated the potential pitfall of the same code being repeated in multiple places, which could lead to problems where the code is improved in one place but not everywhere it is used. This is a concept familiar to software programmers as "Don't Repeat Yourself" (DRY).

The task starts with the keyword task and is followed by the task name. The first of these tasks is handle_settings:

```
task handle_settings;
begin
  if (s_up)
  begin
    mins <= mins + 1;
    if (mins == 9)
    begin
      mins <= 1;
    end
  end
  if (s_dn)
  begin
        mins <= mins - 1;
    if (mins == 1)
    begin
      mins <= 9;
    end
  end
end
endtask
```

This task deals with the "Up" and "Down" buttons being pressed to increment and decrement the mins register. It also handles the wrap-around that occurs if you increment the number of minutes over 9 or under 1.

The decrement_time task listed next is responsible for reducing the time by 1 second on every second. It uses the prescaler register to just do something every fifty-millionth tick of CLK. The value 49999999 will match the clock frequency of your FPGA (−1), or your timer will run too fast or too slow, although you are unlikely to notice the one-fifty-millionth of a second.

Having hit the 50 millionth −1 tick, prescaler must be set back to 0. The register secs is decremented, and the cascading set of if statements ensures that the other digits are decremented when the preceding digit hits 0:

```
task decrement_time;
begin
  prescaler <= prescaler + 1;
  if (prescaler == 26'd49999999) // 50 MHz to 1Hz
  begin
    prescaler <= 0;
    secs <= secs - 1;
    if (secs < 1)
    begin
      secs <= 9;
      ten_secs <= ten_secs - 1;
      if (ten_secs < 1)
      begin
        ten_secs <= 5;
        mins <= mins - 1;
      end
    end
  end
end
endtask
```

The last task (reset_time) is the simplest: it just sets the secs and ten_secs registers back to 0 and the mins to the last used value of mins recorded in mins_stored as the timer was started:

```
task reset_time;
begin
  secs <= 0;
  ten_secs <= 0;
  mins <= mins_stored;
end
endtask

endmodule
```

Testing

Make up the project, and deploy it onto your board. You should find that the "Up" and "Down" buttons (SW1 and SW6) on the Elbert 2 increment and decrement the minutes displayed. Press the middle button (SW3 on the Elbert) to start the timer. When the timer reaches 0, the buzzer will sound until you press the "Cancel" button (right button on the Mojo, SW4 on the Elbert 2).

Summary

This project makes a good pattern for any project that is complicated enough to use a state machine. The key is to break the module down into tasks. In Chapter 7, you will learn how to generate pulses to control power using pulse-width modulation (PWM) and to control the positions of servo motors.

7

PWM and Servomotors

The parallel nature of an FPGA makes it well suited to generating pulses, either for controlling power using pulse-width modulation (PWM) or for generating the precise streams of timing pulses needed by servomotors. You could, if you really wanted, generate pulses on every GPIO pin of the FPGA simultaneously.

Pulse-Width Modulation

Figure 7-1 shows how PWM works. If the pulses are short (say, high for just 5 percent of the time), then only a small amount of energy is delivered with each pulse. The longer the pulse, the more energy is supplied to the load. When powering a motor, this will control the speed at which the motor rotates. When driving a LED using PWM, the brightness appears to change. In fact, a LED can turn on and off millions of times per second, so the PWM pulses will become pulses of light. The human eye and brain do the averaging trick for us, making the apparent brightness of the LED vary with the pulse length.

The percentage of time that the PWM output is high is called the *duty cycle* or just *duty*. In practice, rather than use a percentage, it is more common to express the duty as a value between 0 and an upper limit that is a power of 2 minus 1. A very common range is 0 to 255, where 0 is completely off and 255 is on all the time.

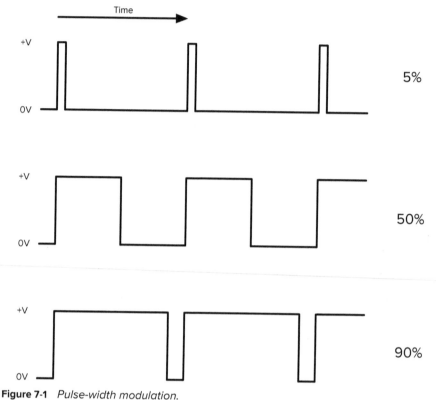

Figure 7-1 *Pulse-width modulation.*

A PWM Module

It is surprisingly easy to implement a PWM module. You use a counter and compare the value of duty with the counter. If the counter is below the duty value, the output is high, and as soon as this isn't true, the output is set low.

You can find the code for this module along with a module to test it in the downloads for this book (see Chapter 2). The project is called "ch07_pwm."

PWM Module Inputs and Outputs

The PWM module is called just "pwm" and has two inputs, pwm_clk and duty. Its single output, PWM_PIN, will be connected to the GPIO pin on which you want the PWM output:

```
module pwm(
    input pwm_clk,
    input [7:0] duty,
    output reg PWM_PIN
    );
```

The `pwm_clk` input is not the same as the system-wide clock of the FPGA. Generally, the PWM frequency is a lot lower than the 12 to 50 MHz of this book's example boards. A common PWM frequency range is 500 Hz to several kilohertz. This PWM module uses an 8-bit counter, so the frequency at `pwm_clk` needs to be 256 times the desired PWM frequency. When you look at testing the module, you will see how you can use a prescaler counter to derive a lower clock frequency to use for the PWM module.

The remainder of the code for the PWM module is as follows:

```
reg [7:0] count = 0;

always @(posedge pwm_clk)
begin
  count <= count + 1;
  PWM_PIN <= (count < duty);
end

endmodule
```

The 8-bit counter count is incremented with the positive edge of the `pwm_clk` signal. Output `PWM_PIN` is then set to be the result of (count < duty). In other words, if count is less than duty, `PWM_PIN` will be 1; otherwise, it will be 0.

A Tester of the PWM Module

In this case, the module to test the PWM module so that you can see that it is working is more complicated than the pwm module itself. It will change the brightness of a LED (D1 on the Elbert 2, LED 0 on the Mojo IO Shield, and LED0 on the Papilio with LogicStart MegaWing) when you press the "Up" and "Down" buttons (SW1 and SW6 on the Elbert 2).

The UCF for Mojo is as follows:

```
# User Constraint File for PWM on Mojo with IO Shield

NET "CLK" LOC = P56;

# Switches
NET "switch_up" LOC = "P137" | PULLDOWN;
NET "switch_dn" LOC = "P139" | PULLDOWN;

# Output
NET "PWM_PIN" LOC = P97; # LED1
```

The tester module has inputs of the system clock (CLK) and two switch pins to vary the brightness of the LED up and down. The output will link to the LED whose brightness is to be changed. The code for the tester module is contained in the file pwm_tester.v:

```
module pwm_tester(
    input CLK,
    input switch_up,
    input switch_dn,
    output PWM_PIN
    );
```

Wires and debouncer modules are defined in what should now be a familiar way:

```
wire s_up, s_dn;
debouncer d1(.CLK (CLK), .switch_input (switch_up), .trans_up (s_up));
debouncer d2(.CLK (CLK), .switch_input (switch_dn), .trans_up (s_dn));
```

A register duty is used to keep a value of duty cycle (0 to 255) that will be set using the "Up" and "Down" push switches.

The prescaler register is a 7-bit counter that will be used to divide the system clock frequency (50 MHz on the Mojo, 12 MHz on the Elbert 2, and 32 MHz on Papilio One) by 128. With a further division of 256 from the 8-bit counter in the PWM module, this will result in PWM frequencies of:

- Elbert 2: 12 MHz/128/256 = 366 Hz.

- Papilio One: 32 MHz/128/256 = 975 Hz

- Mojo: 50 MHz/128/256 = 1.53 kHz

The pwm module instance uses the output of bit 6 of the prescaler to provide the pwm_clk input to the PWM module:

```
reg [7:0] duty = 0;
reg [6:0] prescaler = 0; // CLK freq / 128 / 256 = 1.5kHz
pwm p(.pwm_clk (prescaler[6]), .duty (duty), .PWM_PIN (PWM_PIN));
```

The always block of the tester module increments the prescaler and then checks for any switch presses. The "Up" and "Down" switch presses increase or decrease the value of duty by 5:

```
always @(posedge CLK)
begin
  prescaler <= prescaler + 1;
  if (s_up)
  begin
    duty <= duty + 5;
  end
  if (s_dn)
  begin
    duty <= duty - 5;
  end
end

endmodule
```

To keep the Verilog simple, there is no test to see if the value of duty exceeds 255. If it does, it will simply wrap around because only 8 bits are used for the duty, allowing a maximum range of 255.

Trying It Out

Generate the bit file for the project, and load it onto your board. You will see how the "Up" and "Down" buttons increase and decrease the brightness of the LED. If you want to get a feel for how PWM works, try slowing the whole thing down by a factor of approximately 1000. To do this, change lines 13 and 14 of pwm_tester.v to add another 10 stages to the prescaler counter. The changes are highlighted in boldface here:

```
reg [16:0] prescaler = 0;

pwm p(.pwm_clk (prescaler[16]), .duty (duty), .PWM_PIN (PWM_PIN));
```

Servomotors

Servomotors (Figure 7-2) are a combination of motor, gearbox, and sensor that are often found in remote-control vehicles to control steering or the angles of surfaces on remote-control airplanes and helicopters.

Unless they are special-purpose servomotors, they do not rotate continuously. They usually only rotate through about 180 degrees, but they can be accurately set to any position by sending a stream of pulses. Figure 7-3 shows a servomotor and shows how the lengths of the pulses determine the position of the servo.

Figure 7-2 *A small 9-g servomotor (left) and standard RC servomotor (right).*

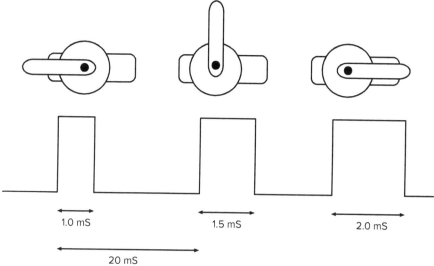

Figure 7-3 *Controlling a servomotor with pulses.*

A servomotor has three connections: GND, a positive power supply (5 to 6 V), and a control connection. The GND connection is usually connected to a brown or black lead, the positive connection to a red lead, and the control connection to an orange or yellow lead.

Although the motor can draw considerable current, the control connection draws very little current. The servomotor expects to receive a pulse every 20 ms or so. If the pulse is 1.5 ms in duration, then the servo will sit at its middle position. If the pulse is shorter, then it will settle in a position to one side, and if the pulse is longer, it will move to a position on the other side of the center position.

Hardware

To control a servomotor with your FPGA board, you are going to need a few parts.

You Will Need

In addition to your FPGA board you are going to need the following items to build this example project:

Part	Source
Mini 9-g servomotor	Adafruit, product ID 196
3× Female-to-male jumper leads	Adafruit, product ID 1953
Male-to-male jumper lead	Adafruit, product ID 760
DC barrel-jack-to-screw-terminal adapter	Adafruit, product ID 368
6-V battery box or 6-V at 1-A power supply	Adafruit, product ID 830

Although it is possible to power at least a small servomotor from the 5-V supply of your FPGA, it is better to use a separate power supply because the large startup currents can cause the FPGA to reset.

Construction

Figure 7-4 shows the project wired up on the Mojo board. Neither the Mojo nor the Papilio needs to have a shield attached. The following connections should be made:

- Negative supply lead of the servomotor (black or brown) to a GND connection on the FPGA using a male-to-male jumper wire

Figure 7-4 *A servomotor controlled by a Mojo (using external power supply).*

- Negative connection from the screw terminal adapter to another GND connection on the FPGA

- Control lead of the servomotor (orange or yellow) to P31 (header P1 pin 1) on the Elbert 2 or P97 on the Mojo IO or P31 (header C pin 0) on the Papilio One using a male-to-male jumper wire

- Positive lead of the servomotor to the positive terminal on the screw terminal adapter

If you have a small servomotor and want to see if it will work from the Mojo or the Papilio's 5-V power connection, you can connect the positive-supply connection lead on the servomotor directly to the board and avoid using an external power supply at all. This is shown in Figure 7-5. Note that the Elbert 2 does not have an accessible 5-V terminal.

Figures 7-6 and 7-7 show a servomotor wired to an Elbert 2 and Papilio One, respectively.

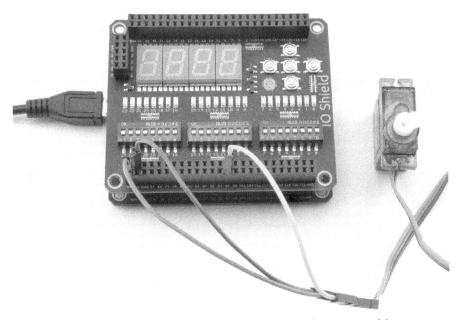

Figure 7-5 *A servomotor controlled by a Mojo (no external power supply).*

Figure 7-6 *A servomotor controlled by an Elbert 2 board.*

Figure 7-7 *A servomotor controlled by a Papilio One board.*

A Servo Module

The servo module and a test program to use it can be found in the project "ch07_servo." The servo module itself is in the file servo.v and is actually the same for all three boards despite their different clock speeds.

The module has inputs of CLK and pulse_len. The input pulse_len is the duration of the pulse in microseconds. This is a 16-bit number giving a maximum duration of 65,536 μs (65 ms). Given that the pulses are going to be arriving every 20 ms, this is actually far more than is needed for the servo's requirement of a pulse between 1 and 2 ms. This allows you to use this module for high-resolution PWM.

The modules output CONTROL_PIN will carry the train of pulses:

```
module servo(
    input CLK,
    input [15:0] pulse_len,   // microseconds
    output reg CONTROL_PIN
    );
```

To get around the problem of timing being messed up by the differing clock speeds of the FPGA boards, a *parameter* is used. This differs from the input and output parameters of the module in that it is evaluated when the project is being synthesized (the bit file generated) rather than when it's actually running. You declare a parameter like this:

```
parameter CLK_F = 50; // CLK freq in MHz
```

Two 16-bit registers, `prescaler` and `counter`, are used to control the pulse timing:

```
reg [15:0] prescaler;
reg [15:0] count = 0;
```

The `always` block is synchronized with the FPGA's clock, and the first step is to add 1 to the prescaler. The prescaler will reach the value of the parameter `CLK_F - 1` every microsecond. When it does so, the prescaler will be reset, and then 1 will be added to `count`. The control pin is then set to the result of comparing `count` with `pulse_len` in the same way as the PWM code at the start of this chapter.

To maintain a 20-ms pulse length, when `count` reaches 19,999, it is also reset:

```
always @(posedge CLK)
begin
  prescaler <= prescaler + 1;
  if (prescaler == CLK_F - 1)
  begin
    prescaler <= 0;
    count <= count + 1;
    CONTROL_PIN <= (count < pulse_len);
    if (count == 19999) // 20 milliseconds
    begin
        count <= 0;
    end
  end
end


endmodule
```

There are two variations of the test code. The Elbert 2 and Mojo use "Up" and "Down" buttons (SW3 and SW5 on Elbert 2) and a joystick on the Mojo IO Shield to nudge the pulse length (and servo position) up or down by 100 ms. The Papilio's LogicStart shield covers up the GPIO pin connectors, so the LogicStart is not used, which means no buttons to control the servo. So the test program for this board cycles the pulse length up in steps of 100 μs until it gets to 2,500 μs, and then it resets it to 500 μs.

Both test programs provide pulses in the range 0.5 to 2.5 ms, which is a slightly wider range than the standard servo pulse range, so you will find that there might be some jitter in the servos at the ends of the range. Using the wider range does allow you to use the maximum range of rotation for your servomotor.

The test program for the Elbert 2 (servo_tester.v) is as follows:

```verilog
Module servo_tester(
    input CLK,
    input switch_up,
    input switch_dn,
    output CONTROL_PIN
    );


wire s_up, s_dn;
debouncer d1(.CLK (CLK), .switch_input (switch_up), .trans_up (s_up));
debouncer d2(.CLK (CLK), .switch_input (switch_dn), .trans_up (s_dn));

reg [15:0] pulse_len = 500; // microseconds

servo #(12) p(.CLK (CLK), .pulse_len (pulse_len), .CONTROL_PIN
        (CONTROL_PIN));

always @(posedge CLK)
begin
  if (s_up)
  begin
    pulse_len <= pulse_len + 100;
  end
  if (s_dn)
  begin
    pulse_len <= pulse_len - 100;
  end
end

endmodule
```

The test program is very similar in operation to the test program for PWM. The main thing to note is that where the servo module is instantiated, the text #(12) is used to override the CLK_F parameter of the servo module to the value 12. If you had multiple parameters, the values would be separated by commas.

The code for the Mojo version is almost identical, but the `always` block for the Papilio version that steps the servo from one end of its travel to the other is as follows:

```
always @(posedge CLK)
begin
  prescaler <= prescaler + 1;
  if (prescaler == 32000000) // 1Hz
  begin
    pulse_len <= pulse_len + 100;
    if (pulse_len == 2500)
    begin
      pulse_len <= 500;
    end
  end
end
```

It uses its own prescaler to add 100 to `pulse_len` every second until it reaches 2,500.

Summary

Chapter 8 will turn your attention from controlling power using PWM and servomotors to generating pulses to make sounds. This includes revisiting the alarm module that you made use of in the countdown timer project of Chapter 6.

8

Audio

The Mojo, Papilio, and Elbert 2 both have built-in 3.5-mm sockets designed to attach a pair of headphones or an audio jack to an external amplifier. It is also pretty straightforward to connect a GPIO pin of the Mojo to an audio amplifier. In this chapter, you will start by looking at the "alarm" module that you used in the countdown timer of Chapter 6 and then go on to look at some more complex sound-generation examples including playing a short recorded sound file.

Simple Tone Generation

The code for the "alarm" module used in Chapter 6 is listed next for the Papilio One and LogicStart shield. You can find the code for this in both the "ch06_count-down_timer" and "ch08_alarm" projects in the downloads for this book (see Chapter 2). Here is the version for the Papilio One (32-MHz clock).

```
module alarm(
    input CLK,
    input enable,
    output reg BUZZER
    );

reg [25:0] count;

always @(posedge CLK)
begin
  count <= count + 1;
  if ((count == 26'd32000) & enable) // 1kHz
  begin
```

```
    BUZZER <= ~ BUZZER;
    count <= 0;
  end
end

endmodule
```

The module generates a 1-kHz signal by using a counter that is incremented on every tick of the CLK. When the counter reaches 32,000, the output BUZZER is toggled, and the counter is reset to 0. Thus, 32,000 ticks at the 32 MHz of the Papilio clock equates to the output tone of 1 kHz. When using this on the 50-MHz-clocked Mojo, the value 50,000 is used, and on the Elbert 2's 12-MHz clock, 12,000 is used.

The following UCF for the Papilio outputs the tone on P41, which is connected to the audio output jack. Plug in a pair of headphones or some powered speakers, and you will hear a fairly ugly 1-kHz square-wave tone. On the Papilio, the note is silenced by pressing slide switch 0, and on the Elbert 2, SW1 silences it. Mercifully, on the Mojo, the tone is only sounded when the "Center" button is pressed.

```
# User Constraint File for 1kHz Alarm tone playing on Papilio
# and LogicStart

NET "CLK" LOC = P89; #32MHz

# Switches
NET "enable" LOC = "P91"; # Slide switch 0

# Output
NET "BUZZER" LOC = "P41";
```

The Elbert 2 version also uses the audio jack, but for the Mojo version, you will need to provide your own connection to powered speakers (see next section).

If you want to use a piezoelectric sounder rather than powered speakers, see Chapter 6.

Audio Output from the Mojo

Figure 8-1 shows the Mojo with IO Shield wired for sound. A 1-kΩ resistor and a short length of solid-core wire are used to connect GPIO pin P97 and GND to an audio lead connected to a powered speaker. The wire and resistor leads are just wrapped around the audio jack's ground and tip segments of the jack plug.

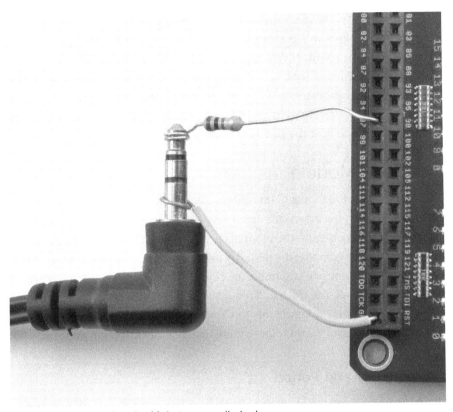

Figure 8-1 *Connecting the Mojo to an audio jack.*

You may want to make something more substantial using some header pins and an in-line 3.5-mm socket.

A General-Purpose Tone/Frequency Generator

The "alarm" project is about as simple as tone generation gets. To improve it, we can make a general-purpose tone-generator module that will be parameterized for the clock frequency of your board and will also allow you to specify the tone to generate as one of its inputs.

You can find the project directory for this module in the downloads for this book in the directory ch08_tone. See Chapter 2 for instructions on downloading the code for this book.

Rather than play the generated frequency through the audio jack, the test program for this module will generate three different frequencies on three different pins of the FPGA board. If you have an oscilloscope or multimeter with a frequency setting, then you will be able to verify the signal being generated. If you don't, then change the location (LOC) of one of the "tone" NETs to your audio jack as per the "alarm" project and listen to the tones through your amplifier—although you may struggle to hear the 12.5-kHz tone.

The Tone Module

The tone module has inputs of the system clock (CLK) and the period of the tone to be generated in microseconds and a single output of tone_out. A parameter CLK_F is used to configure the module's prescaler to suit the clock frequency of your board. The reason that the period is used rather than the frequency is that to convert the frequency to a number of clock cycles to count in Verilog would require division (see the sidebar "Frequency and Period"). Division by anything except powers of 2 (1, 2, 4, 8, 16, etc.) in Verilog is not possible without the use of a division module and would require more than one clock cycle to perform.

Frequency and Period

The *frequency* of a clock is the number of complete cycles (0 to 1 back to 0 again) per second. The *period* of a clock is the time taken for one cycle. So, for a very slow clock of 1 Hz (1 cycle per second), the period is 1 second. For a 1-MHz clock, the period is 1/1,000,000 second or 1 microsecond.

For a particular frequency *f*, the period is 1/*f*. The periods listed in Table 8-1 represent some noteworthy frequencies that you might like to use with the tone module.

Frequency (Hz)	Period (microseconds)	Notes
20	50,000	Lower limit of human hearing
20,000	50	Upper limit of human hearing
1,000	1,000	
261.63	3,822	Middle C

Table 8-1   *Noteworthy Frequencies and Periods*

```
module tone(
    input CLK,
    input[31:0] period, // microseconds
    output reg tone_out
    );

parameter CLK_F = 32; // CLK freq in MHz
```

Two counters are used. The prescaler counter reduces the clock frequency to 2 MHz, and the counter counter is used to count the prescaled clock. The prescaler generates a 2-MHz clock rather than 1 MHz because the output is going to be toggled (0 to 1 or 1 to 0) each time the correct period is reached, effectively halving the frequency again. This toggling ensures that the signal generated has a 50 percent duty cycle. That is, the square wave is high and low in equal amounts.

```
reg [5:0] prescaler = 0;
reg [31:0] counter = 0;

always @(posedge CLK)
begin
  prescaler <= prescaler + 1;
  if (prescaler == CLK_F / 2 - 1)
  begin
    prescaler <= 0;
    counter <= counter + 1;
    if (counter == period - 1)
    begin
      counter <= 0;
      tone_out <= ~ tone_out;
    end
  end
end

endmodule
```

The tone_tester Module

The tone_tester module creates three instances of the tone module, each on a different pin and at a different frequency. The period_12khz instance will actually produce a frequency of 12.5 kHz rather than 12 kHz.

```
module tone_tester(
    input CLK,
    output tone_1khz,
    output tone_100hz,
    output tone_12khz
    );

reg [31:0] period_1khz = 1000;
reg [31:0] period_100hz = 10000;
reg [31:0] period_12khz = 80;

tone #(32) t1(.CLK (CLK), .period (period_1khz),
                          .tone_out (tone_1khz));
tone #(32) t2(.CLK (CLK), .period (period_100hz),
                          .tone_out (tone_100hz));
tone #(32) t3(.CLK (CLK), .period (period_12khz),
                          .tone_out (tone_12khz));

endmodule
```

Testing

Figure 8-2 identifies the GPIO pins used as outputs on the Papilio One board, and Figure 8-3 shows the same on the Mojo.

The pins used on the Elbert 2 are

- 1-kHz P31 (connector P1, pin 1)

- 100-Hz P32 (connector P1, pin 2)

- 12.5-kHz P28 (connector P1, pin 3)

You will find more about the GPIO pin allocations in the Appendices B to D. Connecting a dual-channel scope to the pins shows the waveforms of Figure 8-4.

Playing an Audio File

This project uses a FPGA to play recorded audio data back through an amplifier. It introduces some useful new techniques including the use of random access memory (RAM) and how to load it with a set of data during synthesis.

You can find the project directory for this module in the downloads for this book in the directory ch08_wav_player. See Chapter 2 for instructions on downloading the code for this book.

Figure 8-2 *Generating multiple frequencies on the Papilio One.*

Figure 8-3 *Generating multiple frequencies on the Mojo.*

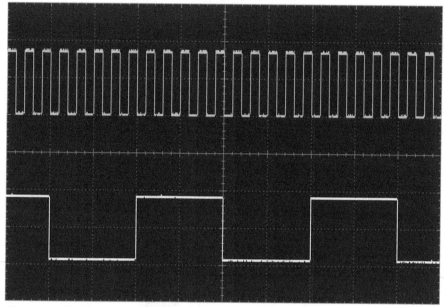

Figure 8-4 *Oscilloscope trace for 1 kHz (top) and 100 Hz (bottom).*

Audio Files

There are many types of audio files, and most use clever compression algorithms to reduce the file size as much as possible with as little loss of quality as possible. At its simplest, an audio file will just contain a long series of numbers, each number representing the amplitude (think voltage) at an instant in time. This format is called a *raw format* because nothing fancy has been done to the numbers. Figure 8-5 shows the data for a sample of me saying the word *one*.

The numeric values at each sample point are held in a single byte (8 bits), giving a range of numbers between 0 and 255. The sound was sampled at 8 kHz, so the approximately 3,800 samples represent less than half a second of audio.

A software tool like Audacity (www.audacityteam.org/) will allow you to record your own audio file or import an audio file from almost any format and then save the result as raw data. Later on, in the section "Preparing Your Own Sounds," you will learn how to use Audacity and a small Python script to convert the raw sound file into a format suitable for importing into a FPGA.

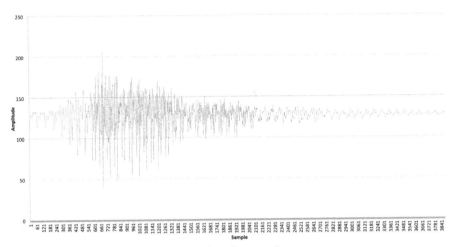

Figure 8-5   *The waveform for the "one" audio sample.*

RAM

The sound to be played is going to be held in memory (RAM) on the FPGA. All the FPGAs used on the three example boards (in fact, most FPGAs) have a dedicated area of the FPGA chip that is used for RAM. When ISE spots a `reg` declaration in your Verilog that looks like memory, it uses this specialized block rather than general-purpose cells. None of the FPGAs used in this book have much RAM. The Elbert 2 has just 54 kbits, the Papilio One has 74 kbits, and the Mojo has most with 576 kbits. This means that even on the Mojo, with a sample rate of 8 kHz, you will only be able to store 6 or 7 seconds of audio. On the Elbert, you will only have about half a second. Halve the sample rate, and you will get double the time, but with a loss of quality.

The RAM will have a number of input address lines and output data lines. You set a number in binary on the address lines to specify the memory byte to which you want access. You can then read or write the currently selected byte. Change the address, and you get access to a different byte of data. Although the whole point of RAM is that you can both read from and write to it, in this project the RAM is filled with initial contents from a file during synthesis and then left unchanged.

The wav_player Module

To play the sound file, a counter is used to step through each address in turn. The value at each location is then output using PWM. The module has inputs of CLK and switch_play, which is used to trigger the sound file to be played when a switch is pressed:

```
module wav_player(
    input CLK,
    input switch_play,
    output reg audio_out
    );
```

Next, there is a local parameter to hold the number of bytes of data contained in the audio file. The format for specifying the memory looks like a regular reg declaration except that as well as specifying the size of each element of the data (7:0) the quantity of such memory locations is also specified (MEM_SIZE-1:0) after the name of the memory (memory):

```
localparam MEM_SIZE = 19783;

reg [7:0] memory[MEM_SIZE-1:0];
```

To load up the memory with data, a special command $readmemh ("read memory hex") is used contained in an initial block. The $readmemh command takes two parameters, the name of the data file to be loaded into the memory and the name given to the memory. This loading happens during synthesis:

```
initial begin
  $readmemh("01_03_b19783.txt", memory);
end
```

The file format for $readmemh is hexadecimal numbers, one per line. In the section "Preparing Your Own Sounds," you will learn how to create such a file. The switch is linked to a debouncer in the usual way:

```
wire s_start;
debouncer d1(.CLK (CLK), .switch_input (switch_play),
            .trans_up (s_start));
```

To control the playing of the sound file, a register play is used. This acts as a flip-flop to turn the pulses to the address counter on and off. The prescaler counter is used to decrease the clock frequency of the board to match the sample rate of the board. The 8-bit counter is used as the PWM counter, and value is used for the contents of the memory currently referred to by address.

Figure 8-6 shows how all these things relate to each other. It also gives a clue as to what the Verilog is likely to be synthesized to.

```
reg play = 0;
reg [3:0] prescaler;
reg [7:0] counter;
reg [19:0] address;
reg [7:0] value;
```

The always block increments the prescaler counter only if play is 1. If you skip down to the end of the always block, you will see that this happens when the "Start" button is pressed. Assuming that the button has been pressed, the prescaler does its job of dividing the clock frequency from the 50-MHz input clock (this is the Mojo version) down to 2 MHz. Then counter is incremented, and the data from the current memory location are latched into value. The value of the current memory location is then compared with counter and audio_out set appropriately.

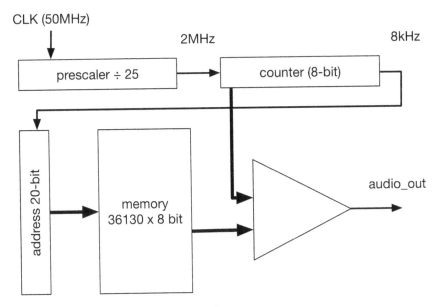

Figure 8-6 *Schematic for the sound file player.*

This results in a PWM output with pulses that are longer for a high amplitude and shorter for a low amplitude. When the value of address reaches MEMORY_SIZE, play is reset to stop the prescaler clocking and prevent any further audio playing until the button is pressed again:

```
always @(posedge CLK)
begin
  if (play)
  begin
    prescaler <= prescaler + 1;
    if (prescaler == 15)  // 8kHz x 256 steps = 2.048 MHz
    begin
      prescaler <= 0;
      counter <= counter + 1;
      value <= memory[address];
      audio_out <= (value > counter);
      if (counter == 255)
      begin
        address <= address + 1;
        if (address == MEM_SIZE)
        begin
          play <= 0;
          address <= 0;
        end
      end
    end
  end
  if (s_start)
  begin
    play <= 1;
  end
end
endmodule
```

Testing

The Elbert 2 and Papilio (with LogicStart MegaWing) will both play the audio through their audio sockets. To play audio from the Mojo and IO Shield, connect it up as you did in Figure 8-1. When you press the "Start" button, you should hear the sound file being played.

Preparing Your Own Sounds

Preparing your own sound file is pretty straightforward. The main problem is keeping it short enough to fit into RAM. Start by installing and running Audacity (www.audacityteam.org/). Audacity is available for Windows, Mac, and Linux and is free. Start a new project, and make sure that the "Project Rate" is set to 8 kHz and that the recording is set to mono (Figure 8-7). Make a short recording by clicking on the circular red "Record" button.

When you have finished recording, press the "Stop" button (yellow square), and you will see the waveform of what you have recorded. If the waveform is a little flat, you can either record it again but talk louder or amplify the signal digitally by selecting the Effect → Ampilfy ... option on the menu. You will also probably want to trim off any silence from the start and end of the recording, which you can do by selecting the part of the waveform that you want to remove and then pressing the DELETE key.

You now need to export the sound file in a very particular format, so select the menu option File → Export Audio.... You need to select a file type of "Other Uncompressed" and then in the header field select "RAW" (headerless) and in the encoding field select "Unsigned 8-bit PCM." Then hit "Save" to create the file (Figure 8-8).

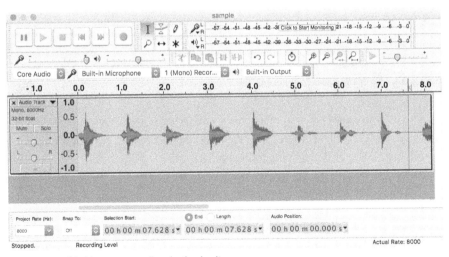

Figure 8-7   *Making a recording in Audacity.*

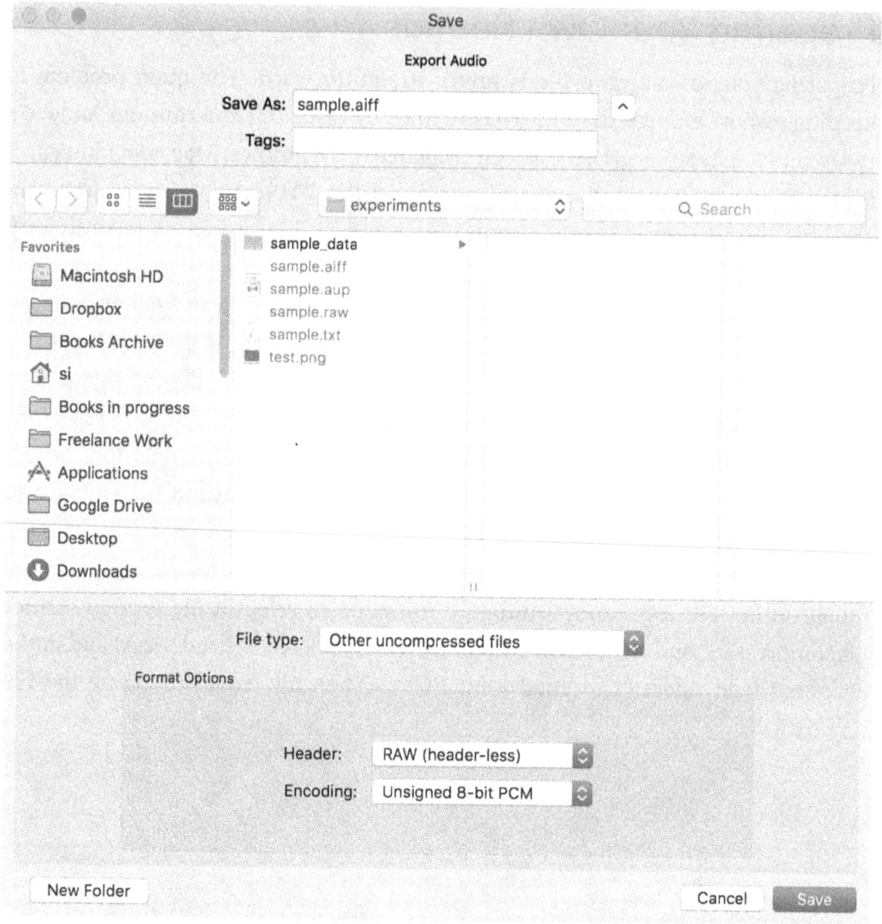

Figure 8-8 *Exporting the raw sound file.*

The file you have just created is binary, and to put it into a form suitable for the $readmemh file, it needs to be converted into a text file with two-digit hexadecimal strings, one per line like the following example:

```
82
89
8e
89
```

To do this conversion, you can use the Python script that I have created for this purpose, which you will find with the downloads for this book in the folder "utilities/audio." Before you can run this script, you will need to install Python, following the instructions at https://wiki.python.org/moin/BeginnersGuide/Download. The program will work in Python 2 or Python 3.

Put the sound file you exported into the same folder as the script ("utilities/audio"), and then run the program from the command line using the command:

```
$ python raw2hex.py 01_05.aiff 01_05.txt
Read (bytes)36130
```

The first parameter (in this case, `01_05.aiff`) is the name of the sound file to convert, and the second parameter (`01_05.txt`) is the output text file. The utility helpfully tells you the size of the data generated so that you can use it to set `MEM_SIZE` in `wav_player` before you build the project.

Summary

Generating sounds is all about dividing the clock signal and then generating pulses to produce a sound. In Chapter 9 you will also be generating pulses, but this time for video signals.

9

Video

Most of modern electronics, whether its generating digital sounds, controlling power to a LED, or positioning a servomotor, seems to be concerned with generating pulses. This is also true for the generation of video signals.

Both the Elbert 2 and the Papilio LogicStart have GPIO connectors and a little extra electronics attached to them for generating VGA signals for use with a computer monitor or TV. The projects in this chapter are just for the Elbert 2 and Papilio, but if you have VGA hardware for your Mojo, you should not find it hard to adjust the timings of the code to work with that board too.

VGA

Video Graphics Array (VGA) is an old standard for video connections that has been around since the early days of the PC. It's relatively easy to generate the pulses for it, and most monitors and TVs still have a VGA connection. VGA signals are made up of frames. A *frame* is one complete set of visible pixels plus some invisible timing signals that we will address shortly. The screen is refreshed many times a second, for the examples in this chapter, in the case of the Elbert 2, 60 frames a second and for the Papilio 73 frames a second. This is also often called the *refresh frequency* expressed in hertz. Figure 9-1 shows the VGA hardware for the Papilio LogicStart MegaWing.

If you look at the VGA connector (labeled "VGA" in Figure 9-1), you can see that apart from GND, there are just five connections that are active. These are RED, GREEN, BLUE, VSYNC, and HSYNC.

UGA

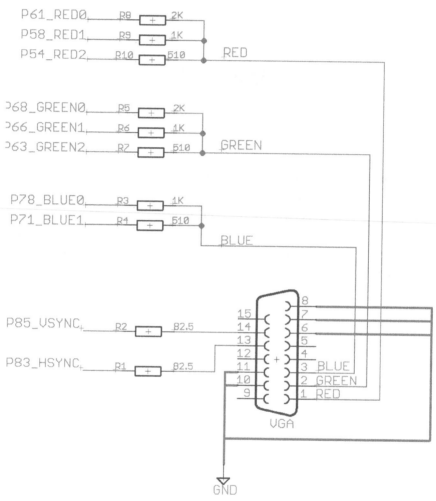

Figure 9-1 *Papilio LogicStart MegaWing VGA schematic.*

The three color signals are analog. That is, the level of each of the signals controls the intensity of that color on the screen. The RED and GREEN channels each use three digital outputs with a simple resistor-based digital-to-analog converter that produces an output intensity in eight steps. The BLUE signal uses just two digital outputs so that it has four possible levels. BLUE loses out so that the entire

RGB color can be specified in a single byte of 8 bits. When, say, the RED and GREEN are both at their highest level for a particular pixel, they will combine by the process of *additive color mixing* (https://en.wikipedia.org/wiki/Additive_color) to produce yellow.

The HSYNC signal is a negative-going pulse that marks the end of each line, and the VSYNC signal marks the end of the whole frame. The timings of HSYNC and VSYNC have some extra conditions of how long the pulses are and when they happen, although many monitors are quite tolerant of deviations from the standard timing. Figure 9-2 shows some of the naming conventions around the pulse generation, with time flowing left to right and top to bottom.

As well as the RED, GREEN, and BLUE signals that control the color of each of the 640 × 480 pixels, a large portion of the "frame" of signals will not correspond to visible pixels but rather to timing signals. To visualize this, it is useful to think of each frame as being bigger than just its visible part in order to contain the extra timing signals needed. That is, there needs to be a horizontal sync (HS) pulse before the start of each full line of pixels and after all the lines have been displayed, and there needs to be a vertical sync (VS) pulse to tell the monitor that it should start displaying a new frame.

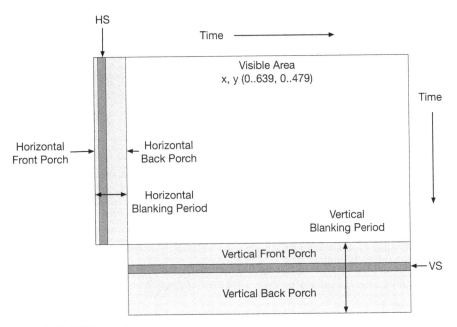

Figure 9-2 *VGA signals.*

So, before the color intensities are written for a particular row of pixels, the HS signal must be generated. This is a negative-going pulse of a fixed duration. There must be a pause called the *horizontal front porch* before the HS pulse and then a longer pause afterwards (the *horizontal back porch*) before the color pulses start to cause the pixels to be displayed. The whole period of the horizontal front porch, HS, and the horizontal back porch is termed the *horizontal blanking period*, and during this period, RED, GREEN, and BLUE signals should all be set to zero.

A similar thing happens with the vertical sync (VS) signal after the full 480 lines of the video have been displayed. This is also surrounded by front and back porches and the whole period termed the *vertical blanking period*. The RED, GREEN, and BLUE signals also need to be zero during this period.

During the visible portion of the scan, the RED, GREEN, and BLUE signals do not need to be sent as separate pulses unless the color changes from one pixel to the next. To set the entire screen red, you just need to keep the three RED control pins high all the time apart from the porches. This means that even though nominally there are 640 pixels across each line, the line will go red whenever the RED signal is high and will stop being red when it goes low again. Physically, the display will be made up of actual pixels at a certain resolution, so matching the ON/OFF times of the colors to the real pixels on the screen surface will produce the best results.

VGA Timings

There are many different options for scan rates and resolutions for VGA, but the two that you will be using in this chapter are listed in Table 9-1. You can find a more complete list of timings at http://web.mit.edu/6.111/www/s2004/NEWKIT/vga.shtml.

Looking at the first row for 640 × 480 pixels at 73 Hz, the first column tells us that the pixel clock needs to be 31.5 MHz. This means that to display the whole frame and also generate the sync signals, a 31.5-MHz clock is needed. During the visible part of the frame, one pixel will be drawn for each tick of the clock. From now on, our main unit of time will be a *tick*, 1 tick being the time period to display a single pixel. This relates to a period of 31.74 ns per pixel (1/31.6 MHz). The horizontal front porch is specified to be 24 ticks, the sync signal needs to go low for 40 ticks, and the back porch needs to be 128 ticks.

	640 × 480 73-Hz refresh (Papilio)	640 × 48060-Hz refresh (Elbert 2)
Pixel Clock (Actual)	31.5 MHz (32 MHz)	25 MHz (24 MHz)
Pixel Tick	31.74 ns	40 ns
Horizontal Front Porch	24 ticks	16 ticks
Horizontal Sync	40 ticks	96 ticks
Horizontal Back Porch	128 ticks	48 ticks
Horizontal Visible	640 ticks	640 ticks
Vertical Front Porch	9 lines	11 lines
Vertical Sync	3 lines	2 lines
Vertical Back Porch	28 lines	31 lines

Table 9-1 *VGA Timings*

When it comes to the vertical part of the signal (which is much longer than the horizontal sync signals), it makes sense to use whole lines as the unit of time, a *line* being the time to generate the horizontal timing signal (porches + HS). So, the vertical front porch is specified as 9 lines, the sync pulse 3 lines, and the back porch 28 lines. These periods are all quite long and reflect the fact that the standard was defined in the days when monitors used cathode-ray tubes (CRTs), and they needed the time to re-aim the electron beam.

The Papilio has a clock frequency of 32 MHz, which is close enough to 31.5 MHz that most modern monitors will be able to adjust themselves to the slightly higher frequency. Hence for the Papilio, you will use the 73-Hz settings. The Elbert 2 has a 12-MHz signal. At first sight, this would seem to rule out its use in generating VGA signals of any kind. However, if we are prepared to make each pixel 2 ticks wide rather than 1, then we can use the 60-Hz VGA standard with an equivalent clock speed of 24 MHz.

Drawing Rectangles

One of the simplest things you can do with a VGA project is to display a few colored rectangles on the screen, as shown in Figure 9-3. You can find this project in the downloads for this book (see Chapter 2). The project is called "09_vga_basic."

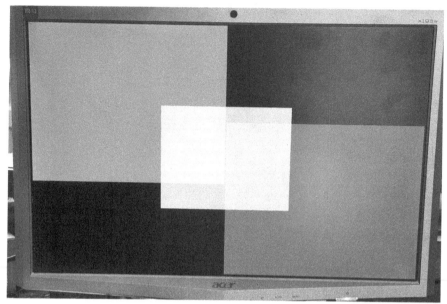

Figure 9-3 *Colored rectangles on the screen.*

A VGA Module

To make our VGA experiments as modular as possible, you will start by creating a module (vga.v) that generates the sync signals. This section starts with the Papilio, and we will look at the Elbert later.

The module has a single input (CLK) and outputs for HS (HS) and VS (VS) that will be wired to the VGA socket. The module also has 10-bit outputs for the current *x* and *y* coordinates. These will also include nonvisible areas of the screen, so the output blank is also provided, and blank will be HIGH if the current scan position is in an area of the frame that is not visible.

The x output will be incremented every pixel tick until the line is complete and then reset. The y output will increment as the signal for each line is completed:

```
module vga(
    input CLK,  // Papilio 32MHz
    output HS, VS,
    output [9:0] x,
    output reg [9:0] y,
    output blank
    );
```

The 10-bit register xc is used to count the x position but including the invisible areas, whereas x is the actual pixel location, so you can use x and y to refer to the actual screen coordinates as you would expect to find them. No equivalent of xc for the y coordinates is needed because the VS signal appears after all the visible lines have been displayed:

```
reg [9:0] xc;
```

The following section of assignment blocks is used to generate the HS, VS, blank, and x signals based on the scan position of xc and y. You will see how xc and y count when we get to the always block.

If xc is smaller than 192 (the horizontal blanking period) or greater than 832 pixels of width (visible pixels + horizontal blanking period) or y is more than 479 (after the last visible pixel), then the blank output will be high. Similarly, the HS signal is calculated using the position of xc. If xc is between 24 and 64 pixels, then it's the right time for HS to be active. Note the use of ~ (not) to invert the signal because HS is active low. The VS signal is generated in the same way.

The x output is calculated as being xc minus 192. The special : ? syntax is shorthand for an if statement. The condition comes first (xc < 192), and if that is true, the value after the question mark (?) is used (0); otherwise, the value xc − 192 is used. This prevents the value of x ever wrapping round, which would happen if the result of the subtraction were to be negative.

```
// Horizontal 640 + HFP 24 + HS 40 + HBP 128 = 832 pixel ticks
// Vertical, 480 + VFP 9 lines + VS 3 lines + VBP 28 lines
assign blank = ((xc < 192) | (xc > 832) | (y > 479));
assign HS = ~ ((xc > 23) & (xc < 65));
assign VS = ~ ((y > 489) & (y < 493));
assign x = ((xc < 192)?0:(xc - 192));
```

The always block follows the pattern we have seen many times, where two counters (xc and y) are used to keep track of the current scan position of the frame including the invisible parts:

```
always @(posedge CLK)
begin
  if (xc == 832)
  begin
    xc <= 0;
    y <= y + 1;
```

```
      end
      else begin
        xc <= xc + 1;
      end
      if (y == 520)
      begin
        y <= 0;
      end
  end
end

endmodule
```

The following module (vga_basic.v) tests out the vga module to draw the rect-angles of Figure 9-3. The inputs and outputs to this top-level module connect to the NETs defined in the UCF for the project.

```
module vga_basic(
    input CLK,
    output HS,
    output VS,
    output [2:0] RED,
    output [2:0] GREEN,
    output [1:0] BLUE
    );
```

Two 10-bit wires x and y are linked to the instance of vga. Note that in this example the "blank" output of vga is not needed.

```
wire [9:0] x, y;
vga v(.CLK (CLK), .HS (HS), .VS (VS), .x (x), .y (y));
```

This is where things might suddenly seem backwards from the normal approach to drawing things, where the software programmer would want to set the color at a particular x and y pixel on the screen. What happens here is that instead, the logic for determining the magnitude of each color is expressed in terms of x and y. So for a red square from 1 (1 to 299), 299 can be drawn by specifying that red should be ON if x is greater than 0 and x is less than 300 and y is greater than 0 and y is less than 300. The ?: syntax is used to ensure that RED is set to 7 for maximum red brightness; otherwise, it would be 1. The GREEN and BLUE rectangles are generated in a similar way:

```
assign RED = ((x > 0) & (x < 300) & (y > 0) & (y < 300))?7:0;
assign GREEN = ((x > 200) & (x < 400) & (y > 150) & (y < 350)?7:0);
assign BLUE = ((x > 300) & (x < 600) & (y > 180) & (y < 480))?3:0;

endmodule
```

VGA and the Elbert 2

The 12-MHz Elbert 2 version of vga.v is as follows:

```
module vga(
    input CLK,   // Papilio 32MHz
    output HS, VS,
    output [9:0] x,
    output reg [9:0] y,
    output blank
    );

reg [9:0] xc;

// Horizontal 640 + fp 16 + HS 96 + bp 48 = 800 pixel clocks
// Vertical, 480 + fp 11 lines + VS 2 lines + bp 31 lines = 524 lines
assign blank = ((xc < 160) | (xc > 800) | (y > 479));
assign HS = ~ ((xc > 16) & (xc < 112));
assign VS = ~ ((y > 491) & (y < 494));
assign x = ((xc < 160)?0:(xc - 160));

always @(posedge CLK)
begin
  if (xc == 800)
  begin
    xc <= 0;
    y <= y + 1;
  end
  else begin
    xc <= xc + 2;
  end
  if (y == 524)
  begin
    y <= 0;
  end
end

endmodule
```

This version takes the timings from the second row of Table 9-1 but is otherwise much the same as the Papilio version. The other difference (highlighted in bold) is that xc is increased by 2 each tick rather than 1. This is what gives us the apparent clock speed of 24 MHz by halving the horizontal pixel resolution.

Making Things Move

Figure 9-4 shows a second example. This time, there is a white rectangle drawn around the edge of the screen and a green square that can be moved about the screen using the push buttons on the Elbert 2 or the joystick on the Papilio One and LogicStart.

The vga module is unchanged from the preceding example. It is just the top-level module (vga_game.v in the project "09_vga_game") that is different. The module has some extra inputs for the four push buttons:

```
module vga_game(
    input CLK,
    input up_switch,
    input dn_switch,
    input left_switch,
    input right_switch,
```

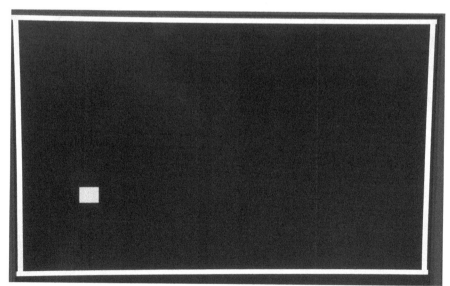

Figure 9-4 *Making things move.*

```
    output HS,
    output VS,
    output [2:0] RED,
    output [2:0] GREEN,
    output [1:0] BLUE
    );
```

This time, in addition to the x and y coordinates, the "blank" signal will be needed to blank out RED, GREEN, and BLUE when not over the visible pixel area. A prescaler is also defined that will slow things down for moving the object about:

```
wire [9:0] x, y;
wire blank;
reg [15:0] prescaler;

vga v(.CLK (CLK), .HS (HS), .VS (VS), .x (x), .y (y), .blank (blank));
```

The position of the object is held in two 10-bit registers, o_x and o_y. The wire "object" 1 or 0 is assigned to the static expression that determines that x and y are inside a square set by the top-right corner o_x, o_y and extending 30 pixels in width and height:

```
reg [9:0] o_x = 320;
reg [9:0] o_y = 240;
wire object = x>o_x & x<o_x+30 & y>o_y & y<o_y+30;
```

A border is also defined in a similar manner. The value of border will be 1 if x and y are at a place where the border should appear:

```
wire border = (x>0 & x<10) | (x>630 & x<640) | (y>0 & y<10) |
              (y>470 & y<480);
```

The values for RED and BLUE are just determined by the border. The value of GREEN is determined by OR-ing together the things contributing to whether GREEN should be ON, that is, whether x and y are in border or are in object:

```
assign RED = (border & ~ blank)?7:0;
assign GREEN = ((border | object) & ~ blank)?7:0;
assign BLUE = (border & ~ blank)?3:0;
```

The always block uses a prescaler to drop the speed of movement to something slow enough to control and then just uses the button presses to adjust the coordinates of the objects o_x and o_y:

```
always @(posedge CLK)
begin
  prescaler <= prescaler + 1;
  if (prescaler == 0)
  begin
    if (~ up_switch)
    begin
      o_y <= o_y - 1;
    end
    if (~ dn_switch)
    begin
      o_y <= o_y + 1;
    end
    if (~ left_switch)
    begin
      o_x <= o_x - 1;
    end
    if (~ right_switch)
    begin
      o_x <= o_x + 1;
    end
  end
end

endmodule
```

A Memory-Mapped Display

A more conventional approach to video display is to memory map it, that is, to use a memory to contain the color of each pixel, and then separate hardware can generate the VGA signals from the memory.

Unfortunately, neither the Elbert 2 or the Papilio have enough memory to allocate a whole byte to all 640 × 480 = 307,200 pixels. But what we can do is to map the memory to bigger (8 × 8) pixels. This will reduce the effective resolution of the display to 80 × 60. You can also load up the memory with an image like you did in the audio file player so that you can display an image on the screen such as the one shown in Figure 9-5.

Figure 9-5 *Displaying an image using memory-mapped VGA output.*

The vga module is unchanged. You will find the top-level module for memory mapping VGA in the file vga_mem.v in the project "09_vga_mem." This starts with the usual imports to connect things up to hardware with the UCF file:

```
module vga_mem(
    input CLK,
    output HS,
    output VS,
    output [2:0] RED,
    output [2:0] GREEN,
    output [1:0] BLUE
    );

wire [9:0] x, y;
```

The memory is 4800 × 8 bits to accommodate the resolution of 80 × 60. The wire mem_index is used to address the memory and color to reference the color at the current pixel position:

```
reg [7:0] mem[4799:0]; // 80 x 60 (8x8 pixels on 640x480)
wire [12:0] mem_index;
wire [7:0] color;
```

The memory is loaded using $readmemh in the same way as the audio file. Later in this section, the process of converting any image into a suitable format will be explained.

```
initial begin
  $readmemh("flag.txt", mem);
end

vga v(.CLK (CLK), .HS (HS), .VS (VS), .x (cx), .y (y),
      .blank (blank));
```

To find the right address in the memory given values of x and y (small pixels), the value of y (0–439) is divided by 8. Verilog can synthesize division hardware, but only for powers of two, because this can be done easily by a process called bit shifting. This value is then multiplied by 80 because there are now 80 pixels across, and the final result is found by adding x divided by 8 to mem_index.

FPGAs do have multiplication hardware, and you do not have to restrict yourself to powers of 2. You may wonder why you can't just multiply y by 10 rather than first dividing it by 8 and then multiplying it by 80. The reason why is that you need to lose what would become the fractional part of dividing y by 8 before then scaling it up to the right row.

The RED, GREEN, and BLUE values are then set from the appropriate bits of "color," with "blank" being used to make sure the colors are not active when the display should be blanked for not being in a visible region:

```
assign mem_index = (y / 8) * 80 + x / 8; // divide y first to lose LSBs
assign color = mem[mem_index];
assign RED = (blank?0:color[7:5]);
assign GREEN = (blank?0:color[4:2]);
assign BLUE = (blank?0:color[1:0]);

endmodule
```

Preparing an Image

In the downloads for this book you will find a folder called "/utilities/video," and in there you will find another Python utility called "image2hex.py" for importing an image, resizing it to 80 × 60, and converting it into a text file suitable for use with $readmemh.

The utility needs Python to be installed on your computer (see Chapter 8) and also requires the Python Image Library (PIL). To install PIL (after Python has been installed), run the following command from the command line:

```
sudo pip install pillow
```

Like its audio file equivalent, the utility takes two parameters: the file to convert, which can be a JPG, GIF, PNG, BMP, and so on, and the name of the output file to be created. The file created should always be the same size because the script automatically resizes the image to 80 × 60 pixels.

Summary

This chapter will get you started on the road to generating video from your FPGA board. It's not too much of a stretch to see how a digital oscilloscope with VGA output could be built if the FPGA board has suitable signal-capture hardware attached to it. The site FPGA4fun has implemented an entire pong game using VGA generated from a FPGA board (www.fpga4fun.com/PongGame.html). However, without the use of a processor core on the FPGA, it soon becomes difficult to implement game play using pure hardware.

10

What Next

This is an admittedly short book for what is a complex device. I hope that I have managed to at least achieve the goal of this book's subtitle and get you started with Verilog.

There is plenty more to look at, and this is just the start of your journey into FPGAs. In this chapter, you will see a few other features of ISE and also learn about other ways of programming the Mojo and Papilio boards.

Simulation

Development boards such as the Elbert 2, the Mojo, and the Papilio allow you to try out your FPGA code for real, but ISE also includes a feature that allows you to simulate the behavior of your Verilog in software without actually installing it on a FPGA. The process involves creating a test module that sends stimuli to the module under test and will then generate simulation results like those shown in Figure 10-1.

Under the Hood

Although not especially useful, if you are curious about how the synthesis of Verilog code works, you can take a look at its intermediate results. If you select your top-level module and then unfold some of the options in the "Processes" area (bottom-left quarter of Figure 10-2), you will see an option called "View Technology Schematic." Click on this, and you will get a nice symbolic representation of your top-level module as a block with inputs and outputs. However, if you

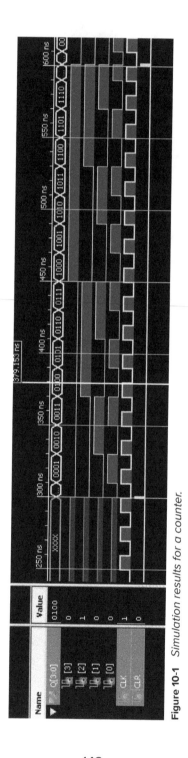

Figure 10-1 *Simulation results for a counter.*

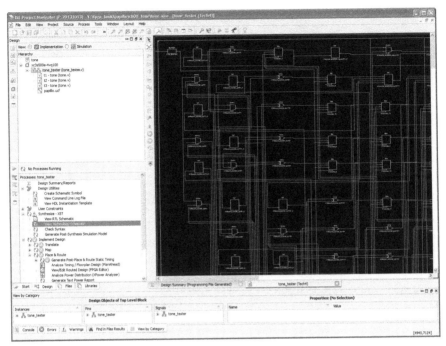

Figure 10-2   *The technology schematic for the "tone" project.*

click on "View Technology Schematic," it will open up to show you all the logic gates used in your synthesis. Figure 10-2 shows just the top left of this schematic for the "tone" project.

Of more use is the "Design Summary." This tells you just how much of your FPGA's resources are being used. When you run the "Generate Programming File" process, you will find that the report will have been added to the documents in the editor area, and you can view it just by clicking on the "Design Summary" tab. Figure 10-3 shows part of the summary for the "countdown_time" project.

Cores and Soft Processors

There is little point in reinventing the wheel, so within ISE and elsewhere, you will find *cores* that can be downloaded and used in your projects. Some of these cores are proprietary and require payment, but there are a growing number of open-source cores and Verilog available under open-source licenses.

Device Utilization Summary				
Logic Utilization	Used	Available	Utilization	
Number of Slice Flip Flops	188	1,408	13%	
Number of 4 input LUTs	168	1,408	11%	
Number of occupied Slices	173	704	24%	
Number of Slices containing only related logic	173	173	100%	
Number of Slices containing unrelated logic	0	173	0%	
Total Number of 4 input LUTs	309	1,408	21%	
Number used as logic	164			
Number used as a route-thru	141			
Number used as Shift registers	4			
Number of bonded IOBs	18	108	16%	
Number of BUFGMUXs	1	24	4%	
Average Fanout of Non-Clock Nets	2.79			

Figure 10-3   *The Design Summary–Device Utilization.*

Be warned, though, that although there is a plentiful supply of open-source Verilog code available to use in your projects, often it is not well documented and lacks examples of using the module, so you will generally have to do quite a lot of work understanding how to use the module.

Both the Papilio and the Mojo have their own *integrated development environments* (IDEs) that you can use as alternatives to ISE. In both cases, they are simpler and easier to use and offer some advantages if you just want to work on one of those boards. However, both also require you to install ISE, which they use transparently to actually do the synthesis.

More on the Papilio

The Papilio IDE is based on the IDE for the popular microcontroller board Arduino and is called the *Papilio Design Lab*. Incidentally, the female header connections of the Papilio One are also compatible with plug-in Arduino shields. The Papilio Design Lab (Figure 10-4) allows you to set up the FPGA on your Papilio One to include a soft processor that is Arduino compatible called the *ZPUino*. You can then add extra components to FPGA design (e.g., VGA output) and then program the ZPUino soft processor as if it were a real processor all from within the Papilio IDE.

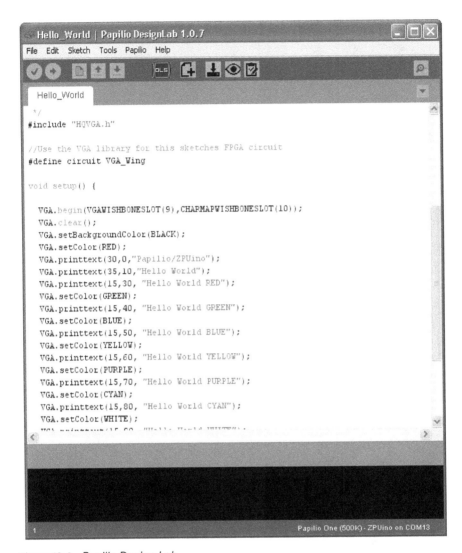

```
Hello_World | Papilio DesignLab 1.0.7

File   Edit   Sketch   Tools   Papilio   Help

Hello_World

*/
#include "HQVGA.h"

//Use the VGA library for this sketches FPGA circuit
#define circuit VGA_Wing

void setup() {

  VGA.begin(VGAWISHBONESLOT(9),CHARMAPWISHBONESLOT(10));
  VGA.clear();
  VGA.setBackgroundColor(BLACK);
  VGA.setColor(RED);
  VGA.printtext(30,0,"Papilio/ZPUino");
  VGA.printtext(35,10,"Hello World");
  VGA.printtext(15,30, "Hello World RED");
  VGA.setColor(GREEN);
  VGA.printtext(15,40, "Hello World GREEN");
  VGA.setColor(BLUE);
  VGA.printtext(15,50, "Hello World BLUE");
  VGA.setColor(YELLOW);
  VGA.printtext(15,60, "Hello World YELLOW");
  VGA.setColor(PURPLE);
  VGA.printtext(15,70, "Hello World PURPLE");
  VGA.setColor(CYAN);
  VGA.printtext(15,80, "Hello World CYAN");
  VGA.setColor(WHITE);
```

Papilio One (500K) - ZPUino on COM13

Figure 10-4 *Papilio Design Lab.*

There are lots of predefined hardware configurations (*circuits*) whose .bit files can just be sent to the FPGA board before its ZPUino is programmed with Arduino C. Creating your own circuits is more complex, requiring the use of ISE.

More on the Mojo

The Mojo's IDE (Figure 10-5) is called simply the *Mojo IDE*, and it encourages the use of the company's HDL called *Lucid*. This is very similar to Verilog with mostly cosmetic changes such as using { and } in place of begin and end.

The IDE is easy to use and includes a collection of *components* that can be added to your design like cores. The Mojo actually has an Arduino-compatible microcontroller attached to it that can provide analog inputs and communication features.

Summary

This book should have started you on your journey with FPGAs or at least shown you their strengths and weaknesses. You will find many resources on the Internet that stir the imagination for that first project to make. As always when learning something new, take small steps, and try a few simple projects before you set out to make something truly ambitious.

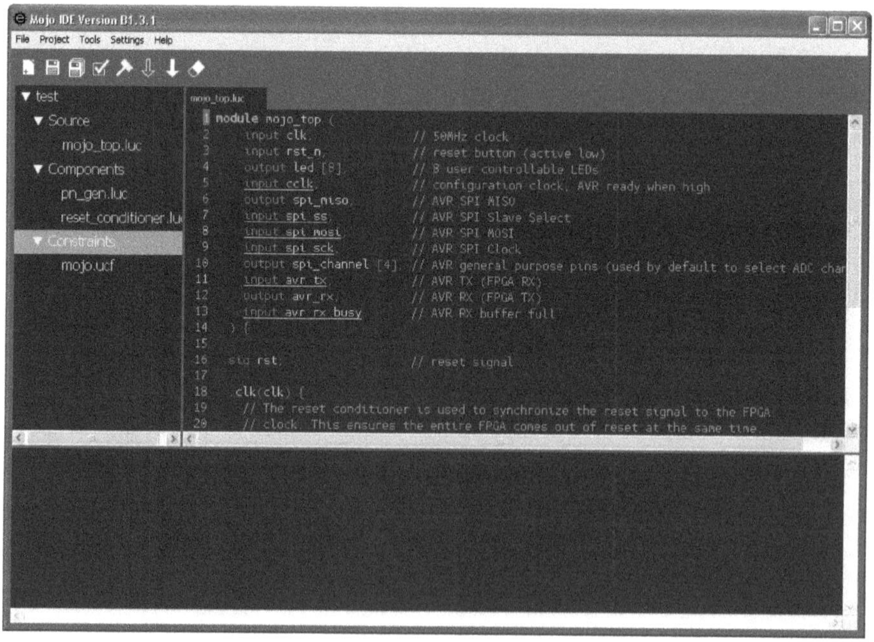

Figure 10-5   *The Mojo IDE.*

Appendix A will point you in the direction of Internet resources to help you in this quest as well as listing some alternatives to the boards used here and places to buy boards and other kit. Appendices B, C, and D provide the detailed information you need about the hardware of the three boards and their interface shields in a consistant structure for quick reference.

Resources

Buying FPGA Boards

The best starting point for obtaining one of the FPGA boards used in this book is the manufacturers' websites:

- **Elbert 2: http://numato.com/.** You can buy from Numato's website, but you can also find the Elbert 2 widely available on Amazon sites around the world.
- **Mojo: https://embeddedmicro.com.** Also available from Adafruit and Sparkfun.
- **Papilio: http://papilio.cc/.** Also available from Sparkfun.

Components

This book is mostly about software, so there are not many components to buy. Each project that needs parts includes a parts list. Some other useful suppliers of components that ship worldwide include

- **Sparkfun:** www.sparkfun.com
- **Adafruit:** www.adafruit.com
- **Mouser:** www.mouser.com
- **Digikey:** www.digikey.com
- **Newark:** www.newark.com

In the United Kingdom, there are

- **Maplins:** www.maplins.com
- **CPC:** http://cpc.farnell.com/

Other FPGA Boards

Many FPGA boards are available for the Xilinx FPGAs that use ISE as well as from manufacturers that have their own tools. In fact, FPGA manufacturers will often also sell "evaluation" boards. These vary considerably in price and are aimed more at professionals rather than "makers."

One intereresting project that aims to get away from the bloated development environments of the FPGA manufacturers is icoBOARD (Figure A-1). This board is being developed as a companion board to the Raspberry Pi and uses an open-source toolchain to synthesize the design. At the time of this writing, the board is in its beta-test phase.

Figure A-1 *The icoBOARD.*

Figure A-2   *The Mimas V2 Spartan 6 board.*

One of the best-value higher-powered FPGA boards is the Numato Labs'
Mimas V2 (Figure A-2). This is essentially an Elbert 2 but with a Spartan 6 FPGA
like the Mojo. The interface hardware (switches, VGA, and audio) are much the
same as those on the Elbert 2.

Web Resources

In addition to the manufacturers' websites mentioned earlier, here are some
resources that you might find useful:

- **fpga4fun.com:** Some great tutorials and Verilog code for all sorts of
  projects and activities

- **opencores.org:** A repository of open-source FPGA projects

- **www.xilinx.com/training/fpga-tutorials.htm:** Tutorials for Xilinx FPGAs
  using ISE (mostly advanced)

- **http://numato.com/learning-fpga-and-verilog-a-beginners-guide
  -part-1-introduction:** A nice FPGA tutorial from Numato Labs

B

Elbert 2 Reference

ISE New Project Settings

When creating a new project in ISE using the New Project Wizard, you should use the settings in Table B-1.

Setting	Value
Evaluation development board	None specified
Product category	All
Family	Spartan3A and Spartan3AN
Device	XC3S50A
Package	TQ144
Speed	−4

Table B-1 *Elbert 2 Project Settings*

Prototype Net Mapping

The Elbert 2's built-in switches, LEDs, and other connectors are summarized in the following sections. All figures are reproduced with permission from Numato Labs.

LEDs

The Elbert 2's individual LEDs are mapped to FPGA pins as shown in Table B-2. The LEDs are connected through series resistors to ground, so a high signal on a LED pin will light the LED.

IO Shield Name (NET)	FPGA Pin (LOC)
D1	P55
D2	P54
D3	P51
D4	P50
D5	P49
D6	P48
D7	P47
D8	P46

Table B-2   *The Elbert 2 LED Pin Locations*

Three-Digit Display

Figure B-1 shows the schemtic for the Elbert 2's seven-segment LED display. The three common anodes of the display are switched by PNP transistors connected to FPGA pins P120, P121, and P124. The digit selection is inverted, so a low at P120 will enable the digit S3.

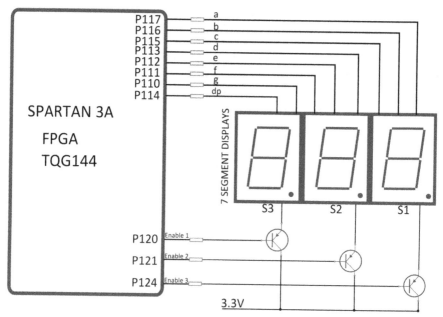

Figure B-1   *Elbert 2 LED display schematic.*

The individual LED segments are controlled by the pins listed in Table B-3.

Segment	FPGA Pin (LOC)
a	P117
b	P116
c	P115
d	P113
e	P112
f	P111
g	P110
DP	P114

Table B-3 *Elbert 2 LED Segment Mapping*

DIP Slide Switches

The Elbert 2's slide switches are mapped to FPGA pins as shown in Table B-4. All the slide switches switch to ground when in the ON position.

Switch	FPGA Pin (LOC)
1	P58
2	P59
3	P60
4	P63
5	P64
6	P68
7	P69
8	P70

Table B-4 *Elbert 2 Slide-Switch Mapping*

Push Switches

The Elbert 2's push switches are mapped to FPGA pins as shown in Table B-5. Each switch is switched to ground, so a pressed switch will make its FPGA pin LOW. There are no pull-up resistors; the inputs are floating.

Switch	FPGA Pin (LOC)
SW1 (top left)	P80
SW2	P79
SW3	P78
SW4	P77
SW5	P76
SW6	P75

Table B-5 *Elbert 2 Slide-Switch Mapping*

VGA

Figure B-2 shows the VGA connector pin mapping. The three channel outputs use a resistor-based digital-to-analog converter (DAC). This provides 3 bits for the red and green channels and 2 bits for the blue channel.

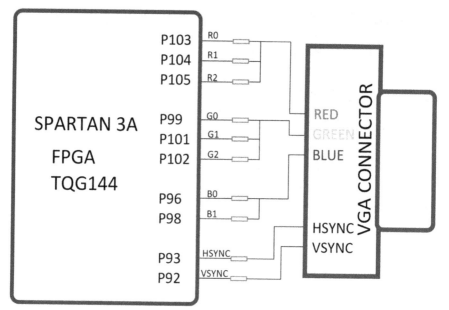

Figure B-2 *Elbert VGA connector mapping.*

Audio and Micro-SD

The Elbert has a 3.5-mm audio socket that can be used to connect the Elbert to an external amplifier. Figure B-3 shows the stereo audio and micro-SD pin mapping. The audio output does not have a low-pass filter.

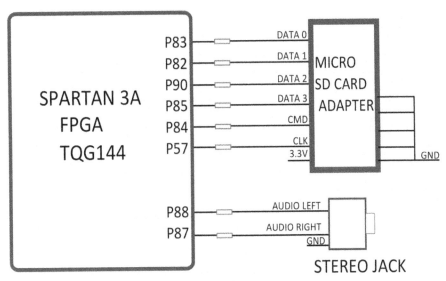

Figure B-3   *Elbert audio socket and SD card.*

GPIO Pins

The following GPIO pins are availble from four headers that are confusingly labeled P1, P6, P2, and P4. Each GPIO pin is 3.3-V logic capable of sourcing or sinking 24 mA.

Header P1

Header Pin (Physical Location)	FPGA Pin (LOC) or Other
1	P31
2	P32
3	P28
4	P30
5	P27
6	P29
7	P24
8	P25
9	GND
10	GND
11	VCC
12	VCC

Header P6

Header Pin (Physical Location)	FPGA Pin (LOC) or Other
1	P19
2	P21
3	P18
4	P20
5	P15
6	P16
7	P12
8	P13
9	GND
10	GND
11	VCC
12	VCC

Header P2

Header Pin (Physical Location)	FPGA Pin (LOC) or Other
1	P10
2	P11
3	P7
4	P8
5	P3
6	P5
7	P4
8	P6
9	GND
10	GND
11	VCC
12	VCC

Header P4

Header Pin (Physical Location)	FPGA Pin (LOC) or Other
1	P141
2	P143
3	P138
4	P139
5	P134
6	P135
7	P130
8	P132
9	GND
10	GND
11	VCC
12	VCC

Clock

The Elbert 2 has a 12-MHz clock on pin P129.

C

Mojo Reference

ISE New Project Settings

When creating a new project in ISE using the New Project Wizard, you should use the settings in Table C-1.

Setting	Value
Evaluation development board	None specified
Product category	All
Family	Spartan6
Device	XC6SLX9
Package	TQG144
Speed	−2

Table C-1 *ISE Settings for the Mojo*

NET Mapping (IO Shield)

The following subsections detail the connections between the FPGA's GPIO pins and the peripherals on the IO Shield.

LEDs

The IO Shield's individual LEDs across the center of the board are mapped to FPGA pins as shown in Table C-2. The LEDs are connected through series resistors to ground, so a HIGH signal on a LED pin will light the LED.

161

IO Shield Name (NET)	FPGA Pin (LOC)
LED 0	P97
LED 1	P98
LED 2	P94
LED 3	P95
LED 4	P92
LED 5	P93
LED 6	P87
LED 7	P88
LED 8	P84
LED 9	P85
LED 10	P82
LED 11	P83
LED 12	P80
LED 13	P81
LED 14	P11
LED 15	P14
LED 16	P15
LED 17	P16
LED 18	P17
LED 19	P21
LED 20	P22
LED 21	P23
LED 22	P24
LED 23	P26

Table C-2   *Mojo IO Shield LED Pin Mapping*

Four-Digit Display

Figure C-1 (reproduced by kind permission of EmbeddedMicro) shows the section of the IO Shield's schematic that deals with the four-digit, seven-segment LED display. Each of the four common-anode displays is enabled using the FPGA pins P9, P10, P7, and P12 from right to left.

The individual LED segments are controlled by the pins listed in Table C-3.

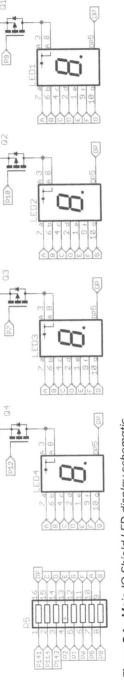

Figure C-1 *Mojo IO Shield LED display schematic.*

163

Segment	FPGA Pin (LOC)
a	P5
b	P8
c	P114
d	P143
e	P2
f	P6
g	P1
DP	P141

Table C-3   *Mojo IO Shield LED Segment Mapping*

Slide Switches

The IO Shield's DIP slide switches are mapped to FPGA pins as shown in Table C-4. The slide switches require the GPIO pin to be pulled down.

Switch	FPGA Pin (LOC)
0	P120
1	P121
2	P118
3	P119
4	P116
5	P117
6	P114
7	P115
8	P112
9	P111
10	P105
11	P104
12	P102
13	P101
14	P100
15	P99
16	P79
17	P78
18	P75
19	P74
20	P67
21	P66
22	P58
23	P57

Table C-4   *Mojo IO Shield Slide-Switch Mapping*

Push Buttons

Figure C-2 (reproduced by kind permission of EmbeddedMicro) shows the section of the IO Shield's schematic that deals with the five push buttons.

This might tempt you to think that any inputs will be pulled up to VCC. They won't! The inputs are floating; the resistors are just used to limit current in the event that the buttons are pressed while their FPGA pin is set to a low output.

You can find the full schematic for the IO Shield at https://embeddedmicro .com/media/wysiwyg/io/IO_Shield.pdf.

Clock Pin

The Mojo provides a 50-MHz clock on pin P56.

Complete UCF for IO Shield

A UCF for the entire IO Shield can be downloaded from https://embeddedmicro .com/media/wysiwyg/io/io.ucf.

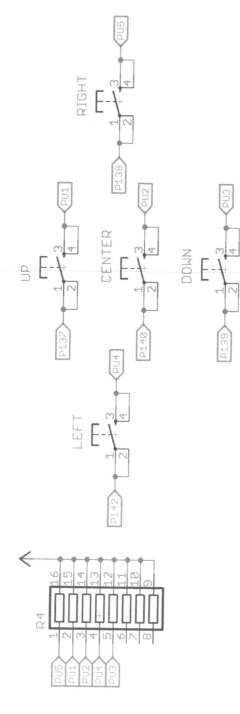

Figure C-2 *Mojo IO Shield push-button schematic.*

166

D

Papilio One Reference

ISE New Project Settings

When creating a new project in ISE using the New Project Wizard, you should use the settings in Table D-1.

Setting	Value
Evaluation development board	None specified
Product category	All
Family	Spartan3E
Device	XC3S250E or XC3S500E[a]
Package	VQ100
Speed	−4

[a]If you have a Papilio One 250, use XC3S250E; for a Papilio One 500, use XC3S500.

Table D-1 *Papilio One Project Settings*

LogicStart MegaWing NET Mapping

The Papilio One does not have any built-in interface devices, so in this book it is assumed that you will be using a LogicStart MegaWing board fitted on top of your Papilio One. You can find the documentation for the MegaWing at http://papilio .cc/index.php?n=Papilio.LogicStartMegaWing.

LEDs

The LogicStart's individual LEDs are mapped to FPGA pins as shown in Table D-2. The LEDs are connected through series resistors to ground, so a high signal on a LED pin will light the LED.

Logic Start Name (NET)	FPGA Pin (LOC)
LED0	P5
LED1	P9
LED2	P10
LED3	P11
LED4	P12
LED5	P15
LED6	P16
LED7	P17

Table D-2 *Papilio Logic Start's LED Pin Locations*

Four-Digit Display

The four common anodes of the display are switched by PNP transistors connected to FPGA pins P67, P60, P26, and P18. The digit selection is inverted, so a low at P67 will enable the digit 0. The design is very similar to that of the Mojo. So, to see a schematic of how the LED display is arranged, see Appendix C. The individual LED segments are controlled by the pins listed in Table D-3.

Segment	FPGA Pin (LOC)
A	P57
B	P65
C	P40
D	P53
E	P33
F	P35
G	P62
DP	P23

Table D-3 *LogicWing LED Segment Mapping*

DIP Slide Switches

The Papilio's slide switches are mapped to FPGA pins as shown in Table D-4. All the slide switches connect to a digital input via a 4.7-kΩ resistor. The switches themselves switch the common contact to either GND or 3.3 V. There are therefore no pull-up resistors.

Switch	FPGA Pin (LOC)
0	P91
1	P92
2	P94
3	P95
4	P98
5	P2
6	P3
7	P4

Table D-4 *LogicWing Slide-Switch Mapping*

Joystick Switches

The LogicWing's joystick push switches are mapped to FPGA pins as shown in Table D-5. Each switch is switched to ground, so a pressed switch will make its FPGA pin LOW. There are no pull-up resistors; the inputs are floating.

Switch	FPGA Pin (LOC)
Select (center push)	P22
Up	P25
Down	P32
Left	P34
Right	P36

Table D-5 *LogicWing Joystick Mapping*

VGA

Figure D-1 shows the VGA connector pin mapping. The three channel outputs use a resistor-based digital-to-analog converter (DAC). This provides 3 bits for the red and green channels and 2 bits for the blue channel.

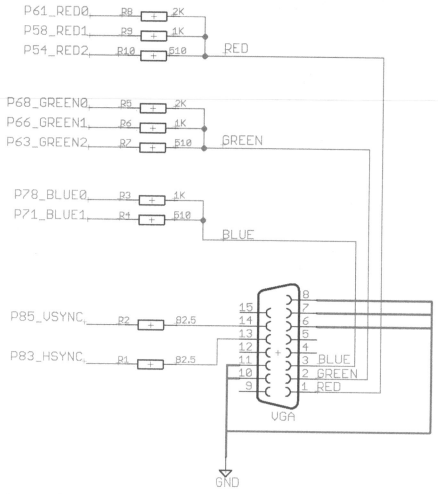

Figure D-1 *LogicWing VGA connector mapping.*

Audio

The LogicStart MegaWing has a 3.5-mm audio socket that can be used to connect the LogicStart MegaWing to an external amplifier. The output is mono using P41, but it does have a low-pass filter.

Analog-to-Digital Converter

An ADC128S102 IC provides eight analog inputs using an SPI interface, as shown in Figure D-2.

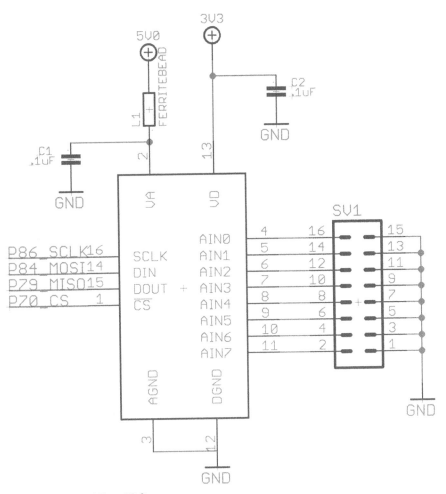

Figure D-2 *LogicWing ADC.*

Clock Pin

The Papilio One provides a 32-MHz clock on P89.

GPIO Pins

Figure D-3 shows the GPIO pin mapping to the Papilio One's connectors.

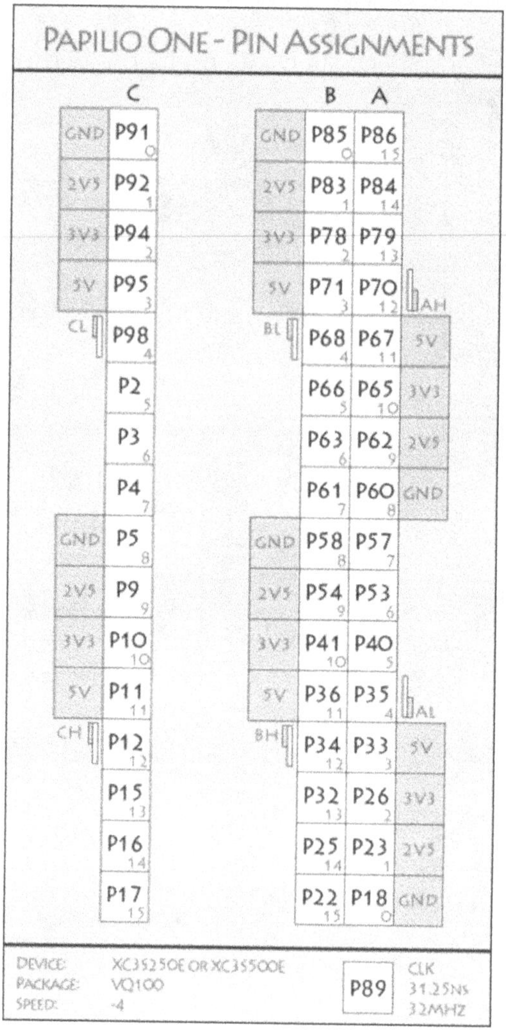

Figure D-3 *Papilio One connectors.*

INDEX

Note: Page numbers in italic refer to figures.

CPSIA information can be obtained
at www.ICGtesting.com
Printed in the USA
FSHW020254171020
74819FS